# 横琴野生鸟类

# THE WILD BIRDS OF HENGQIN

横琴粤澳深度合作区城市规划和建设局　珠海市观鸟协会 ◎ 编

SPM 南方传媒 | 花城出版社

中国 · 广州

图书在版编目（CIP）数据

横琴野生鸟类 / 横琴粤澳深度合作区城市规划和建设局，珠海市观鸟协会编. -- 广州 : 花城出版社，2023.11
ISBN 978-7-5749-0054-7

Ⅰ. ①横… Ⅱ. ①横… ②珠… Ⅲ. ①野生动物－鸟类－介绍－珠海 Ⅳ. ①Q959.708

中国国家版本馆CIP数据核字(2023)第212914号

出 版 人：张 懿
责任编辑：黎 萍 秦翊珊
责任校对：李道学
技术编辑：林佳莹
封面设计：何 成

书　　名　横琴野生鸟类
HENGQIN YESHENG NIAOLEI
出版发行　花城出版社
（广州市环市东路水荫路 11 号）
经　　销　全国新华书店
印　　刷　珠海市豪迈实业有限公司
（珠海市斗门区白蕉镇城东金坑中路 19 号 4 栋厂房二楼）
开　　本　787 毫米×1092 毫米　16 开
印　　张　21.5　2 插页
字　　数　300,000 字
版　　次　2023 年 11 月第 1 版　2023 年 11 月第 1 次印刷
定　　价　168.00 元

如发现印装质量问题，请直接与印刷厂联系调换。
购书热线：020－37604658　37602954
花城出版社网站：http://www.fcph.com.cn

# 《横琴野生鸟类》编委会

委托单位：横琴粤澳深度合作区城市规划和建设局

承担单位：珠海市观鸟协会

总负责人：谭惠芳　罗华兵　郭竣工

主创人员：张　惠　吴杏妍　陈星亮　李　明　吴俊宝　张福庆
郭竣工　陈什旺　崔腾辉　李芳华　汪克宏　乐伟强
吴坎宋　钟　莜　禹　林

参与成员：简丽英　谢　燚　黎潮扬　谭椿议　陈　兰　张健慧
张　凯　邓小龙　张仰曾　叶志明　张俊轩　龚　琴
周明吉　黄成植　詹善衡

技术指导：詹肇泰　廖晓东　王　剑

文字撰写：陈什旺　汪克宏　李芳华

英文翻译：王修志　梁凌紫　李芳华　钟　莜

图片拍摄：张福庆　陈什旺　崔腾辉　郭竣工　汪克宏　朱卫民
乐伟强　柯　大

编校团队：邹峭华　曾丽苏　廖晓东　徐运祥　汪克宏　崔腾辉
梁凌紫　钟　莜

# Editorial Committee of

# *THE WILD BIRDS OF HENGQIN*

# 序言（一）

横琴地处珠海南端，澳门西侧，与珠海和澳门一水相隔，总面积约 106 平方千米，独特的自然地理孕育了森林、海洋、湿地等生态系统和丰富多样的生物资源。

自 2021 年 9 月 17 日横琴粤澳深度合作区挂牌成立起，我局便承担了生态环境保护的相关责任，秉持推动绿色发展、促进人与自然和谐共生的生态文明建设理念，我们积极开展横琴生物多样性本底调查，实施滨海湿地修复、大小横琴山造林、野生动物巡护救助等行动，试点建设国家级湿地公园，举办大湾区生物多样性研讨会，提升生物多样性治理能力，为重点保护物种营造优良的栖息环境，目前在鸟类保护方面的工作成效尤其突出，2021 年至 2023 年 2 月新增野生鸟类记录已高达 64 种。漫步在横琴的海岸沙滩、林荫湖畔甚至市井里巷，总能遇见翱翔长空的黑鸢、踏浪觅食的白鹭、翩跹起舞的白头鹎。

为生动展示横琴野生鸟类保护成果，持续增强公众生态环境保护、推进生态文明建设的行动自觉，我们将调查到的野生鸟类数据及其生态图片汇编成《横琴野生鸟类》一书。书中共收录野生鸟类 222 种，隶属于 18 目 55 科，其中列入国家级重点保护野生鸟类 37 种；除介绍各种鸟类的形态、国家保护等级和生活习性等外，还附录了趣味盎然的野生鸟类观察手记，是了解横琴野生鸟类的最佳科普读物。

希望本书能为更多人提供一个了解野生鸟类的窗口，吸引更多人关注并参与生物多样性保护，加快建设人与自然和谐共生的中国式现代化。

谨向参与本书出版工作的同人和关心横琴自然生态的社会各界致以衷心的感谢！

**张国基**

横琴粤澳深度合作区城市规划和建设局局长

# Preface 1

Nestled at the southern tip of Zhuhai and the western side of Macao, Hengqin stands gracefully separated from its neighboring cities by glistening waters, boasting a generous expanse of approximately 106 square kilometers. Embraced by its distinctive natural geography, this pristine land fosters a variety of ecosystems, including lush forests, expansive oceans, and flourishing wetlands, all teeming with abundant and diverse biological treasures.

Ever since the momentous inauguration of the Guangdong–Macao In–Depth Co–operation Zone in Hengqin on that auspicious day of 17th September 2021, the Bureau has wholeheartedly taken up the mantle of safeguarding ecological environmental protection. Guided by the noble vision of fostering an ecological civilization that champions sustainable growth and harmonious cohabitation with nature, we have zealously embarked on a multitude of endeavors. These include the dynamic execution of the Hengqin Biodiversity Background Survey, the ambitious restoration of coastal wetlands, the reforestation of Dahengqin Mountain and Xiaohengqin Mountain, the vigilant safeguarding of wildlife through dedicated patrols and rescues, and the pioneering efforts in establishing a national wetland park. Furthermore, we have successfully curated profound seminars on biodiversity in the Greater Bay Area, bolstered our capacity in biodiversity management to ensure the welfare of crucially protected species.

Notably, our ongoing efforts in bird conservation have yielded remarkable results, with an exceptional tally of 64 new wild bird species documented in the year 2021. Whether strolling along the picturesque coasts and sandy beaches, traversing the verdant woodlands and tranquil lakeshores, or meandering through the charming alleys of the city, one is invariably enchanted by the sight of black kites majestically soaring in the cerulean skies, elegant egrets gracefully foraging amidst the undulating waves, and light–vented bulbuls playfully twirling to their own melodious tunes.

With the aim of vividly showcasing the accomplishments in wild bird conservation of Hengqin, perpetually fostering public consciousness regarding ecological environmental protection, and the pursuit of an ecological civilization, we meticulously curated the book titled "The Wild Birds of Hengqin". This publication compiles comprehensive data on surveyed wild birds alongside their picturesque photograph. The book encompasses an impressive compilation of 222 species of wild birds, spanning 18 orders and 55 families. Remarkably, 37 species among them are included in the roster of nationally protected key wild birds. The book enriches its content by offering detailed insights into the morphology, national protection level, habits, etc. Furthermore, captivating handwritten notes on the observation of wild birds adorn its pages, rendering it the ultimate popularization resource for comprehending the splendid avian diversity of Hengqin.

This book aspires to be a captivating gateway, enticing numerous individuals to immerse themselves in the captivating world of wild birds. It seeks to ignite a deeper sense of awareness and active commitment to biodiversity conservation, while simultaneously propelling the progress of a Chinese-style modernization that celebrates the harmonious symbiosis between humankind and the natural world.

I extend my sincere gratitude to the colleagues who contributed to the publication of this book and to the community that demonstrates a genuine concern for the natural ecology of Hengqin.

Cheong Kok Kei

Secretary for Urban Planning & Construction Bureau, Guangdong-Macao In-Depth Co-operation Zone in Hengqin

# 序言（二）

## 远眺的横琴

第一次邂逅横琴是在20多年前的20世纪90年代末，那时是澳门回归的前后。当时我在香港特别行政区政府渔农自然护理署工作，负责主管香港特区鸟类保育工作。原本只想去看一下澳门氹仔的湿地，并看一下在那里度冬的黑脸琵鹭，因为当时全球黑脸琵鹭数目不足1000只，被评为极度濒危物种。也因为氹仔那边的土地要发展了，从当时的认知，氹仔是香港—深圳后海湾以外在珠江三角洲有黑脸琵鹭具体记录的地方。欲穷千里目，更上一层楼。我走上澳门塔，站在离地200多米的观光层，360°远眺，蓝天白云，赫然发现在西面的十字门水道（当时不知道叫十字门）背后有一大片渺无人烟的绿水青山，较前的位置有片芦苇和红树，后面是一个个长方形排列有序的大型围垦鱼塘，它的规模和尺度比著名的候鸟天堂——香港米埔基围还要大许多。之后，我从澳门半岛往氹仔朝南方向的桥上走，近了一点远眺横琴，心里有着如同发现新大陆的兴奋，这真是一片美丽的湿地！

## 近观的横琴

2010年，我的工作转移到珠海。我第二次看横琴是在2011年左右，这次是我真真正正在横琴的土地上走，近距离地看横琴，看到的是一个宏图！青山依旧在，但是一个一个的围垦鱼塘正在堆进泥土，新型建筑正在拔地而起，了解到这片土地将要被建设成一个创新且兼具生态文明的示范新区！

## 深入的横琴

10年间，每次踏足横琴都会给我一次又一次邂逅的感觉，黑脸琵鹭、黄嘴白鹭、一群群的野鸭、夜间神秘的猫头鹰和夜鹰、行不得也哥哥（中华鹧鸪）、让人眼花缭乱的鸻和鹬……另一方面，它的发展仍在不断向前推进，芒洲湿地公园和二井湾湿地公园建立起来，新的人为环境和自然生态环境在交织着。为了摸清横琴岛的陆生野生脊椎动物资源，横琴粤澳深度合作区城市规划和建设局连续数年，委托珠海市观鸟协会开展野生动物的巡护和监测工作。珠海市观鸟协会本着科学、严谨、专业的态度，数年如一日地辛苦耕耘，特别在野生鸟类方面收获了丰富的成果，结合历史文献资料，截至 2023 年 2 月，记录到横琴岛有野生鸟类 222 种。

本图册通过实地拍摄的生态照片见证着蜕变中横琴多样的鸟类资源。看本图册让我们欣慰的是，今天这个 100 平方千米多一点的横琴岛，仍然拥有着比较多样化的生境，也许我们会问将来这些那些能否得以延续，然后让我们想到珍惜当下每段天然海岸线、每片湿地、每处森林。丝萝非独生！横琴不会是也不应该是以一个孤立的生态系统存在，它的自然生态和生物多样性依托相邻东西南北的生境与生态景观的连贯性，无可否认横琴岛是粤港澳大湾区西岸的一个生态组成，希望我们留给下一代的不只是本图集，也不单是这些鸟种，而是一个具有生态可持续的横琴。“绿水青山就是金山银山”，衷心期盼横琴能创造经济发展与自然保育并行的新里程碑。

**詹肇泰博士**

北京师范大学－香港浸会大学联合国际学院生命科学系副系主任、副教授

珠海市观鸟协会名誉会长

# Preface 2

## A distant view of Hengqin

I encountered Hengqin for the first time 20 years ago, around the late 1990s, when it was around the handover of Macao. At that time, I was working for the Hong Kong SAR Government's Agriculture, Fisheries, and Conservation Department, and responsible for bird conservation work in the Hong Kong SAR. Initially, I only intended to visit the wetlands of Taipa, Macao, in which the Black-faced Spoonbills were wintering there. At that time, the population of Black-faced Spoonbill was less than a thousand globally, making them critically endangered. Also, the land in Taipa was about to be developed, and based on my knowledge at that time, Taipa was the location where Black-faced Spoonbills was recorded beyond the Hong Kong-Shenzhen Bay. As the saying goes, "To see farther, climb higher.", I went up to the Macao Tower, standing on the observation deck over 200 meters above the ground. From there, I had a panoramic view of the surroundings: blue sky and white clouds, and to my astonishment, I discovered a vast expanse of untouched green mountains and waters behind the Shizimen Canal (which I didn't know it was called Shizimen at the time) in the western area. There were reeds and mangroves in the front, followed by large rectangular fishponds arranged neatly. The scale and size of these fishponds were even larger than the renowned bird paradise, Mai Po Marshes in Hong Kong. Afterwards, as I walked from the Macao Peninsula towards Taipa on a bridge in the southern direction, I got a closer view of Hengqin. My heart was filled with excitement, as if discovering a new continent. It was truly a beautiful wetland!

## A close view of Hengqin

In 2010, I transferred my work to Zhuhai. The second time I saw Hengqin was around 2011, and that time I was actually walking on Hengqin, observing it up close. What I saw was a grand vision! The green mountains were still there, but the fishponds were reclaimed one by one. New buildings were rising from the ground, and I learned that this land was going to be developed into a demonstration new area that combines innovation and ecological civilization!

## Exploring deeper into Hengqin

Over the course of a decade, Hengqin made me feel like encountering it anew when I set foot in it every time: Black-faced Spoonbills, Chinese Egrets, flocks of wild ducks, mysterious owls and nightjars, Chinese Francolins, an array of sandpipers and plovers that dazzled my eyes... What is more, its development continues to progress. Mangzhou Wetland Park and Erjingwan Wetland Park have been established, where artificial environments and natural ecological environments intertwine. In order to have an in-depth understanding of the wild terrestrial vertebrates biodiversity in Hengqin Island, The Urban Planning and Construction Bureau, Guangdong-Macao In-Depth Cooperation Zone in Hengqin has commissioned the Zhuhai Bird Watching Society (ZHBWS) to conduct professional surveys for several years. The outcomes of those surveys have significantly deepened our understandings of the biodiversity in Hengqin, which resulted in 222 bird species recorded in Hengqin by February 2023, including records of the historical documentations.

This book illustrated biodiversity of birds in Hengqin during the course of development by the outcomes of those surveys and on-site photographs of the birds, which employed scientific field survey methods and infrared-triggered camera traps, drones, monoculars, binoculars, and DSLR cameras. To our comfort by reading this book, there is a relatively diverse habitats in Hengqin island with a little over 100 square kilometers. Perhaps we may wonder if these habitat could be sustained in the future, which reminds us to cherish every natural coastline, wetland, and forest in the moment. No plant thrives alone! Hengqin should not and cannot exist as an isolated ecosystem. Its natural ecology and biodiversity depend on the connectivity of habitats and ecological landscapes to the east, west, south, and north. Undeniably, Hengqin Island is an integral part of the ecological composition of the western coast of the Guangdong-Hong Kong-Macao Greater Bay Area. We hope to leave future generations not only with this book, not just with these bird species, but with an ecologically sustainable Hengqin. "Green waters and lush mountains are invaluable assets!" We sincerely hope that Hengqin will create a new milestone where economic development and nature conservation go hand in hand.

**Dr. SiuTai TSIM**
Beijing Normal University-Hong Kong Baptist University United International College
Associate Professor and Associate Head, Department of Life Sciences
Honorary President of Zhuhai Bird Watching Society

# 目录 Contents

# 本书使用说明

科目名
Order and Family

2 雁形目 ANSERIFORMES · 鸭科 Anatidae

3 中文拼音
Chinese Pinyin

4 中文名
Chinese name

5 学名
Scientific name

6 英文名
English Name

luó wén yā

**罗纹鸭**

**Falcated Duck**

*Mareca falcata*

**保护等级：** 三有 / 近危

**居留类型：** 冬候鸟

**体长：** 46 ～ 54 厘米

**生活习性：** 常与其他鸭类混群活动于河口、湖泊、海湾等湿地生境；滤食水中浮游动物、软体动物及植物的根茎等。

**Protection categories：** 3V /NT

**Status：** Winter visitor

**Body length：** 46–54 cm

**Habitat and Habits：** Often found mingling with other duck species in wetland habitats like river estuaries, lakes, and bays. Filter-feed on plankton, mollusks, and the roots and stems of aquatic plants in the water.

雄鸟 ♂

雌鸟 ♀

\>>> 023

1 页码
Page

7 保护等级：
Protection Category
参考本书299页
Refer to P303

8 季节居留情况 Residency Category
留鸟Resident bird
冬候鸟Winter visitor
夏候鸟 Summer visitor
过境鸟Passing visitor

9 体长
Body length

10 生活习性 Habitat and habit
典型的行为方式和栖息环境
Typical behavior and habitat

11

图片说明 Photograph Illustration

1) 雌鸟与雄鸟 Female and Male：♂：雄鸟 Male；♀：雌鸟 Female

该鸟种雌雄异色，且在横琴均拍摄到雄鸟和雌鸟，放两张照片。

该鸟种雌雄异色但在横琴仅拍摄到雄鸟或雌鸟，或该鸟种雌雄同色，放一张照片。

Two photographs for one species: plumage variation of male and female birds of one species and photographs were taken for both in Hengqin.

One photograph for one species:

Case I: plumage variation of male and female birds of one species but photographs were taken for only one sex in Hengqin.

Case II: no plumage variation of male and female birds of one species.

2) 繁殖羽与非繁殖羽 Breeding plumage (br.) and Non-breeding plumage(non-br.):

br.繁殖羽；non-br:非繁殖羽

该鸟种的繁殖羽和非繁殖羽有明显差异，且在横琴拍摄到这两种羽色的个体，放两张照片；

该鸟种的繁殖羽和非繁殖羽有明显差异，但在横琴仅拍摄到某种羽色的个体，或该鸟种的繁殖羽和非繁殖羽无明显差异，则放一张照片。

# 横琴概况

# Hengqin

# 自然地理

### ·地理位置

横琴岛位于珠海市西南部，处于珠江入海口，西临磨刀门水道，东临十字门水道与澳门隔海相望，北与珠海南湾城区隔马骝洲水道相望，濒临南海，在北纬 21° 48′ 到 22° 27′ 与东经 113° 03′ 到 114° 18′ 之间。南北长 8.6 千米，东西宽 7 千米，环岛岸线约 50 千米。

### ·地质

横琴岛的地质主要由花岗岩组成，由花岗岩地质形成的丘陵水量不丰沛，丘陵上植被普遍矮小。横琴岛位于珠江入海口处，受入海泥沙淤积影响，海岸地区多为沙滩和泥滩，泥沙滩受淡咸水影响，底栖生物丰富，丰富的底栖生物养育了横琴岛丰富的脊椎动物。横琴岛地理环境多样，有丘陵、有沙滩、有泥滩，孕育着丰富的生物资源。

### ·地貌

横琴岛的地貌包括山地、丘陵、平地（围垦地、填海地）、滩涂和水面，从形态上可分为北、中、南三个部分。北部为小横琴山及填海地、围垦地，中部为中心沟围垦区及填海地、滩涂，南部为大横琴山及周边丘陵、平地、滩涂和水域。

横琴岛地处珠江入海处，珠江水携带大量泥沙在此淤积，使海岸逐渐升高，岸线向外延伸，形成典型的滨海滩涂地貌。滩涂坡度平缓，由淤泥和淤泥质土组成。

### ·气候气象

横琴岛地处南亚热带，濒临南海，气候温和，阳光充足，雨量充沛，夏秋两季多台风，终年气温较高，偶有阵寒，但冬无严寒，夏不酷热。年月温差小，属南亚热带海洋性季风气候，四季温差变化不大，年平均气温 22.4℃左右，年均日照时数达 2000 小时以上，年降雨量 1800–2200 毫米，雨季旱季分明，全年降雨量的 80% 集中在 4–10 月，11 月至次年 3 月为旱季。夏季盛吹东南风，冬季盛吹东北风。6–10 月多台风，8 级强台风平均每年有一次登陆或影响本区域。

## · 水文

横琴岛四面环海，北部为马骝洲水道，东部为十字门水道，西部为磨刀门水道，南部为南海。马骝洲水道位于珠江水系磨刀门出海口附近，径流作用占绝对优势，河道充填沙、河道两侧指状自然堤和拦门沙沉积发育。西江磨刀门水道是珠江八大出海口中最大的分流、分沙水道，同时，由于年输沙量大，造成磨刀门水道口淤积、浅滩发育。横琴岛的海湾有大横琴湾、二横琴湾、深井湾、长沙栏、大东湾等，大部分港湾内为泥滩，退潮时干出，高潮时水很浅。沿岸除西南、东南及突出部分为陡岩外，大部分为垒石岸，间有部分沙质岸。横琴岛有水库四座，分别是望天台水库、牛角坑水库、红旗村水库、团结水库。

## · 水生动物及植被

横琴岛四周的马骝洲横水道、西江磨刀门水道及临岸浅海水域均有众多的浮游藻类、浮游动物、底栖动物以及鱼类、虾类、蟹类等。

横琴岛地处热带北缘地域，原生植被具有热带雨林性质和特征。陆地生态环境良好，植被主要有马尾松、细叶桉、荔枝、龙眼等。区内多为河口与滨海沉积滩涂荒草地，主要是禾本科植物。

横琴岛四面环水，潮间带生长有红树林。横琴岛的红树林主要分布于南湾、石栏洲、望眉角、二井湾湿地、芒洲湿地、澳门大学周边潮间带。红树林生态系统防风固土的效果非常明显，故被称为海岸的“防护卫士”。红树林生态系统也为各种生物提供栖息地，使之成为世界上生物多样性丰富的自然生态系统之一，秋茄、木榄等胎生红树植物在此繁衍生息，迁徙的鸟类、独特的水生生物等在此觅食栖息，对河口海岸的自然生态平衡起着特殊作用。红树林海岸是热带、亚热带海滨独特的自然景观，具有较高的生态旅游价值。

# Natural Geography

## · Location

Hengqin Island ( 113° 03' to 114° 18' E and 21° 48' to 22° 27' N ) is located in the southwest of Zhuhai, facing to the South China Sea. It is in the estuary area of the Pearl River, next to Modaomen Waterway in the west, Shizimen Waterway in the east, Maliuzhou Waterway in the south, and facing Macao across the sea. The Hengqin island is 8.6 kilometers long from north to south, 7 kilometers wide from east to west, with a coastline of about 50 kilometers.

## · Geology

The geology of Hengqin Island is primarily composed of granite. The granite hilly terrain has limited water resources, resulting in generally stunted vegetation on the hills. Due to the location at the estuary of the Pearl River in the Guangdong–Hong Kong–Macao Greater Bay Area, the coastal areas of Hengqin is affected by the sedimentation of mud and sand from the sea, therefore they are characterized by sandy and muddy beaches. Abundant benthos in the mudflats, which is brought by the brackish water, has nurtured the vertebrates in Hengqin island. The island's diverse geographical environment, including hills, sandy beaches, and mudflats, nurtures a rich variety of biological resources.

## · Landscapes

The main geomorphic types of Hengqin Island include mountains, hills, flat land (reclaimed land), mudflats and waterbodies. Morphologically, the island could be divided into three parts: north, middle and south. The northern part is the Lesser Hengqin Mountain and the reclaimed land; the middle part is the reclaimed land and mudflats; the southern part is the Greater Hengqin Mountain and surrounding hills, flat land, mudflats and waterbodies.

Hengqin Island is located at the estuary of the Pearl River, where the water carries a lot of sediments, causing the coast to gradually rise and the shoreline to extend outward, forming a typical coastal mudflat landform. The mudflat has a gentle sloping and consists of silt and silty soil.

## · Climate and Weather

Hengqin Island is located in the south subtropical zone, near the South China Sea, with a mild climate. There is abundant sunshine and rainfall, with high temperature all year round, with occasional cold spells, but no severe cold in winter and no heat in summer. The annual average temperature is about 22.4 ℃ , with an annual average sunshine hour more than 2000 hours and annual rainfall about 1800~2200 millimeters. The rainy season and dry season are distinct, 80% of the annual rainfall is concentrated between April and October, and the dry season is from November to March. The wind blows southeast in summer and northeast in winter, and there are many typhoons from June to October, with a strong grade-8 typhoon landing or affecting the region once a year on average.

## · Hydrology

Hengqin island is surrounded by waters, with the Maliuzhou waterway to the north, the Shizimen waterway to the east, the Modaomen waterway to the west, and the South China Sea to the south. The Maliuzhou Waterway is located near the estuary of the Pearl River, with the sand-filled watercourse and finger-shaped dikes on both sides of the bund because of sedimentation of sand. The Xijiang River is the largest diversion and sand-sharing waterway among the eight major estuaries of the Pearl River. At the same time, due to the large amount of annual sand transport, the Modaomen Waterway is silted up and shoals are developed.

The bays of Hengqin Island include Large Hengqin Bay, Second Hengqin Bay, Shenjing Bay, Changsha Bay, Dadong Bay, etc. Most of the bays are mudflats, which dry out at low tide and are very shallow at high tide. Along the shore, most of them are rocky bands with some sandy part, except for the southwest, southeast and prominent part of the steep rocks. Hengqin Island has four reservoirs: Wangtiantai Reservoir, Niujiaokeng Reservoir, Hongqi Village Reservoir and Tuanjie Reservoir.

## · Aquatic Animals and Vegetation

There are a lot of planktonic algae, zooplankton, benthic animals, fish, shrimp and crabs in the shallow sea of Hengqin Island, Maliuzhou Waterway and Modaomen Waterway. Most of fishes are brackish and sea fish, anadromous fishes are following, and freshwater fishes are rare.

Hengqin Island is located at the northern edge of the tropic area, the native vegetation has the nature and characteristics of tropical rainforest. The terrestrial ecological environment is good, and the vegetation mainly consists of horsetail pine, fine leaf eucalyptus, lychee, longan and so on. The estuarine and coastal sedimentary mudflats are mainly occupied by graminoid plants.

Hengqin Island is surrounded by water on all sides, and mangroves grow in the intertidal zone. The mangrove forests on Hengqin Island are mainly located in Nanwan, Shilanzhou, Wangmeijiao, Erjingwan Wetland, Mangzhou Wetland, the intertidal zone around the University of Macau. The mangrove ecosystem is highly essential in stabilizing shorelines, preventing erosion and protecting the land from waves and storms, so it is called "Guardian of the Coast". The mangrove ecosystem also provides habitats for various organisms, making it one of the most biodiverse natural ecosystems in the world. Viviparous plants such as Kandelia candel and black mangrove grow here, and migrating birds and unique aquatic organisms feed and inhabit here, playing a special role in the natural ecological balance of the estuarine coast. The mangrove coast is a unique natural landscape of the tropical and subtropical seashore, with high ecological tourism value.

黑翅长脚鹬、林鹬和白鹭 Black-winged Stilt, Wood Sandpiper and Little Egret

# 人文地理

横琴岛毗邻港澳，处于“一国两制”的交会点和“内外辐射”的接合部，地理位置极为优越。2009 年 12 月 16 日，横琴新区管委会在珠海市横琴岛正式成立，是广东省人民政府的代理机构，受珠海市政府的委托管理。横琴岛位于中国乃至东南亚经济区的中心，其地缘优越、区域经济示范作用非常明显。横琴岛四面环水，气候温和，海湾众多，有着保存完好的原始植物，水、农林、旅游资源都特别丰富，为横琴创造了良好的环境。

中共中央、国务院于 2021 年 9 月 5 日正式公布了《横琴粤澳深度合作区建设总体方案》（以下简称《总体方案》），实施范围为横琴岛“一线”和“二线”之间的海关监管区域，总面积约 106 平方千米。其中，横琴与澳门特别行政区之间设为“一线”，横琴与中华人民共和国关境内其他地区之间设为“二线”。《总体方案》明确了横琴粤澳深度合作区战略定位为促进澳门经济适度多元发展的新平台、便利澳门居民生活就业的新空间、丰富“一国两制”实践的新示范以及推动粤港澳大湾区建设的新高地。

根据横琴粤澳深度合作区统计局网站的相关资料，截至 2023 年 5 月，横琴粤澳深度合作区户籍人口 33297 人，比 2008 年末（户籍人口 4229 人）增加了 29068 人，增长了 687.35%。

横琴岛 1992 年被列为 20 世纪 90 年代进一步扩大开放的重点区域，但人口增长极为缓慢，户籍人口从 1996 年的 5313 人到 2010 年的 6913 人，14 年时间只增长了 1600 人，平均每年增长 114 人。随着经济的飞速发展，2010 年后人口增加极为快速，2020 年第七次全国人口普查显示，横琴本岛常住人口为 43618 人，与 2010 年第六次全国人口普查的 6913 人相比，10 年共增加 36705 人，增长 530.86%，年平均增长率为 20.22%。[横琴本岛第七次全国人口普查公报（第一号）——常住人口情况]

# Cultural Geography

Hengqin Island is adjacent to Hong Kong and Macao, and is at the intersection of "one country, two systems" and the bonding area of "internal and external radiation", making it an extremely advantageous location. Hengqin New Area Administrative Committee was officially established in Hengqin Island, Zhuhai at December 16, 2009, as an agency of the People's Government of Guangdong Province and entrusted by the Zhuhai Municipal People's Government. Hengqin Island is located at the center of the economic zone of China and even Southeast Asia. With its superior geographic location, it plays an important role of regional economic demonstration. Hengqin Island is surrounded by water and numerous bays. Mild climate, well-preserved primitive plants, agriculture, forestry, and tourism resources, create a favorable environment.

On Sept 5, 2021, the Communist Party of China (CPC) Central Committee and State Council officially issued The General Plan for Construction of the Guangdong-Macao In-Depth Cooperation Zone in Hengqin. According to The General Plan, the In-Depth Cooperation Zone with the area of 106 square kilometers covering the customs supervised area between the "first line" and "second line" of Hengqin Island. The "first line" refers to Hengqin Island and the Macao Special Administrative Region, while the "second-line" refers to the area between Hengqin Island and other Chinese customs territories. The In-Depth Cooperation Zone will become a new platform to boost the moderate and diversified development of Macao's economy, a new space that provides convenience to Macao residents' life and employment, a new model to enrich the practice of one country, two systems, and a new access to building the Guangdong-Hong Kong-Macao Greater Bay Area.

According to the information from the Statistics Bureau of Guangdong-Macao In-Depth Cooperation Zone in Hengqin, the number of household-registered residents was 33297 by May of 2023, an increase of 29068 or 687.35% over the number of 2008, when it is 4229 household-registered residents.

Since 1992, when Hengqin Island was listed as a key area for further opening up in the 1990s, the population growth has been extremely slow, from 5313 in 1996 to 6913 in 2010, an increase of only 1600 in 14 years, or an average of 114 per year. With the rapid economic development, the population increased extremely fast after 2010. According to the Report of the Seventh National Population Census of Hengqin Island, there were 43618 residents in 2020, an increase of 36705 or 530.86% of the Sixth National Population Census in 2010, which there were 6913.

横琴潟湖栖息鹭群 Flocks of birds in the lagoon of Southern Hengqin

# 鸟类多样性

横琴岛地貌丰富，分别位于南北的大小横琴山自东向西起伏绵延，环岛四周的滨海湿地资源多样，有红树林、芦苇地、泥滩和沙滩。林地与滨海湿地为多种野生动物提供了良好的栖息和觅食环境。

2020 年 7 月至 2023 年 2 月，受横琴粤澳深度合作区城市规划和建设局的委托，珠海市观鸟协会先后承接项目共四个，分别是《珠海市横琴岛陆生野生脊椎动物资源本底调查项目（2020—2021）》《芒洲湿地重点保护野生动物巡护和监测项目（2021—2022）》《横琴粤澳深度合作区环岛南路海滩生物多样性资源调查项目（2022）》《2022 年横琴粤澳深度合作区野生动物（鸟类、两栖和兽类）巡护、监测和科普宣传项目》。协会采取科学的野外样线法和红外自动数码照相等调查方法，运用红外相机、无人机、单筒望远镜、双筒望远镜、单反相机等专业设备，针对不同物种出现的概率设计调查时间和调查频次，收获了颇为丰富的成果，尤其是野生鸟类方面，共记录到 193 种。

结合历史文献资料，截至 2023 年 2 月，横琴岛野生鸟类共 222 种，隶属于 18 目 55 科，其中，非雀形目 128 种，占总数的 58%；雀形目 94 种，占总数的 42%。居留类型为留鸟 77 种、冬候鸟 99 种、旅鸟 35 种、夏候鸟 11 种。

横琴岛野生鸟类中，国家重点保护野生动物 37 种，广东省重点保护野生动物 40 种。

国家一级重点保护野生动物 3 种：黑脸琵鹭、黄嘴白鹭、白腹海雕。

国家二级重点保护野生动物 34 种：花脸鸭、白琵鹭、岩鹭、白斑军舰鸟、鹗、黑翅鸢、凤头蜂鹰、蛇雕、靴隼雕、凤头鹰、赤腹鹰、松雀鹰、鹊鹞、白腹鹞、黑鸢、灰脸鵟鹰、普通鵟、翻石鹬、白腰杓鹬、大滨鹬、阔嘴鹬、褐翅鸦鹃、小鸦鹃、草鸮、领角鸮、栗喉蜂虎、白胸翡翠、红隼、红脚隼、燕隼、游隼、仙八色鸫、画眉、红喉歌鸲。

广东省重点保护野生动物 40 种：罗纹鸭、红胸秋沙鸭、大麻鳽、黄斑苇鳽、栗苇鳽、夜鹭、绿鹭、池鹭、牛背鹭、苍鹭、草鹭、大白鹭、中白鹭、白鹭、黑水鸡、红胸田鸡、白喉斑秧鸡、黑翅长脚鹬、反嘴鹬、中杓鹬、红嘴鸥、黑尾鸥、小黑背银鸥、红嘴巨燕鸥、白额燕鸥、普通燕鸥、灰翅浮鸥、白翅浮鸥、鸥嘴噪鸥、三宝鸟、蓝翡翠、斑鱼狗、寿带、海南蓝仙鹟、白眉鹀、栗耳鹀、小鹀、田鹀、栗鹀、灰头鹀。

横琴岛野生鸟类中，列入世界自然保护联盟（IUCN）濒危物种红色名录近危及以上级别的野生动物 11 种。

IUCN 红色名录濒危（EN）2 种：黑脸琵鹭、大滨鹬。

IUCN 红色名录易危（VU）4 种：黄嘴白鹭、红头潜鸭、仙八色鸫、田鹀。

IUCN 红色名录近危（NT）5 种：罗纹鸭、白腰杓鹬、灰尾漂鹬、红颈滨鹬、弯嘴滨鹬。

值得一提的是，2020 年 7 月—2023 年 2 月横琴开展野生动物调查项目，为横琴岛增加野生鸟类新记录 64 种。

广东省野生鸟类新记录 2 种，即靴隼雕、赭红尾鸲。

珠海市野生鸟类新记录 4 种，包括华南冠纹柳莺、高山短翅蝗莺、栗鹀、红胸秋沙鸭。

横琴岛野生鸟类新记录 58 种，包括花脸鸭、罗纹鸭、红头潜鸭、白琵鹭、大麻鳽、黄嘴白鹭、白斑军舰鸟、凤头蜂鹰、蛇雕、白腹海雕、灰脸𫛭鹰、赤腹鹰、白喉斑秧鸡、白脸鸻、翻石鹬、尖尾滨鹬、长趾滨鹬、丘鹬、针尾沙锥、普通燕鸻、黑尾鸥、小黑背银鸥、白额燕鸥、鸥嘴噪鸥、绿翅金鸠、红翅凤头鹃、中杜鹃、草鸮、普通夜鹰、栗喉蜂虎、戴胜、蚁䴕、燕隼、仙八色鸫、灰喉山椒鸟、赤红山椒鸟、黑枕王鹟、寿带、栗头织叶莺、远东树莺、巨嘴柳莺、冕柳莺、黑眉苇莺、栗颈凤鹛、家八哥、虎斑地鸫、乌灰鸫、白眉鸫、海南蓝仙鹟、蓝歌鸲、红尾歌鸲、红胁蓝尾鸲、黄眉姬鹟、白腹蓝鹟、鸲姬鹟、白眉鹀、栗耳鹀、田鹀。

# Bird Diversity

Hengqin Island boasts diverse geographical features. The Greater Hengqin Mountain and Lesser Hengqin Mountain lying in the south and north, respectively, stretch from east to west. The coastal wetlands surrounding the island are rich in varies types, including mangrove forests, reed beds, mudflats, and sandy beaches. The forest of hills and coastal wetlands provide excellent habitats for a variety of wildlife species.

Between July 2020 and February 2023, the Zhuhai Bird Watching Society (ZHBWS) was commissioned by the Urban Planning & Construction Bureau of Guangdong–Macao In–Depth Cooperation Zone in Hengqin with 4 projects altogether to do surveys of terrestrial vertebrate wildlife resources in Hengqin Island. These 4 projects were Baseline Survey of terrestrial vertebrates in Hengqin (2020–2021), Monitoring of the National Protected Wild Animals in Mangzhou Wetland Park (2021–2022), Survey of Biodiversity Resources of Huandaonan Road in Hengqin (2022), and Monitoring of terrestrial vertebrates in Hengqin (2022). The ZHBWS adopted scientific method, such as line transect and camera–trapping, as well as other survey methods, using infrared cameras, drones, monoculars, binoculars, SLR cameras and other professional equipments. With well–designed survey time and frequency according to the probability of occurrence of different species, excellent results were harvested, especially for wild birds, with altogether 193 species recorded.

Combined with historical literature, 222 species of wild birds were recorded in Hengqin Island up to February 2023, belonging to 18 orders and 55 families. Among them, 128 species (58% of the total) are non–Passeriformes species, and 94 species (42% of the total) are Passeriformes. There are 77 species of resident birds, 99 species of winter visitors, 35 species of passing visitors and 11 species of summer visitors.

Among the wild birds in Hengqin Island, 37 species are listed as the national protected wild animals in China and 40 species are listed as the provincial protected wild animals in Guangdong.

Three species are the national first–class protected wild animals. Black–faced Spoonbill (*Platalea minor*), Chinese Egret (*Egretta eulophotes*), White–bellied Sea Eagle (*Haliaeetus leucogaster*).

Thirty–four species are the national second–class protected wild animals. Baikal Teal (*Sibirionetta Formosa*), Eurasian Spoonbill (*Platalea leucorodia*), Pacific Reef Heron

(*Egretta sacra*), Lesser Frigatebird (*Fregata ariel*), Western Osprey (*Pandion haliaetus*), Black-winged Kite (*Elanus caeruleus*), Crested Honey-buzzard (*Pernis ptilorhynchus*), Crested Serpent Eagle (*Spilornis cheela*), Booted Eagle (*Hieraaetus pennatus*), Crested Goshawk (*Accipiter trivirgatus*), Chinese Sparrowhawk (*Accipiter soloensis*), Besra (*Accipiter virgatus*), Pied Harrier (*Circus melanoleucos*), Eastern Marsh Harrier (*Circus spilonotus*), Black Kite (*Milvus migrans*), Grey-faced Buzzard (*Butastur indicus*), Eastern Buzzard (*Buteo japonicus*), Ruddy Turnstone (*Arenaria interpres*), Eurasian Curlew (*Numenius arquata*), Great Knot (*Calidris tenuirostris*), Broad-billed Sandpiper (*Calidris falcinellus*), Greater Coucal (*Centropus sinensis*), Lesser Coucal (*Centropus bengalensis*), Eastern Grass Owl (*Tyto longimembris*), Collared Scops Owl (*Otus lettia*), Blue-tailed Bee-eater (*Merops philippinus*), White-throated Kingfisher (*Halcyon smyrnensis*), Common Kestrel (*Falco tinnunculus*), Amur Falcon (*Falco amurensis*), Eurasian Hobby (*Falco subbuteo*), Peregrine Falcon (*Falco peregrinus*), Fairy Pitta (*Pitta nympha*), Hwamei (*Garrulax canorus*), Siberian Rubythroat (*Calliope calliope*).

Provincial protected wild animals: 40 species. Falcated Duck (*Mareca falcate*), Red-breasted Merganser (*Mergus serrator*), Great Bittern (*Botaurus stellaris*), Yellow Bittern (*Ixobrychus sinensis*), Cinnamon Bittern (*Ixobrychus cinnamomeus*), Black-crowned Night Heron (*Nycticorax nycticorax*), Striated Heron (*Butorides striata*), Chinese Pond Heron (*Ardeola bacchus*), Eastern Cattle Egret (*Bubulcus coromandus*), Grey Heron (*Ardea cinerea*), Purple Heron (*Ardea purpurea*), Great Egret (*Ardea alba*), Intermediate Egret (*Ardea intermedia*), Little Egret (*Egretta garzetta*), Common Moorhen (*Gallinula chloropus*), Ruddy-breasted Crake (*Zapornia fusca*), Slaty-legged Crake (*Rallina eurizonoides*), Black-winged Stilt (*Himantopus himantopus*), Pied Avocet (*Recurvirostra avosetta*), Eurasian Whimbrel (*Numenius phaeopus*), Black-headed Gull (*Chroicocephalus ridibundus*), Black-tailed Gull (*Larus crassirostris*), Lesser Black-backed Gull (*Larus fuscus*), Caspian Tern (*Hydroprogne caspia*), Little Tern (*Sternula albifrons*), Common Tern (*Sterna hirundo*), Whiskered Tern (*Chlidonias hybrid*), White-winged Tern (*Chlidonias leucopterus*), Gull-billed Tern (*Gelochelidon nilotica*), Oriental Dollarbird (*Eurystomus orientalis*), Black-capped Kingfisher (*Halcyon pileata*), Pied Kingfisher (*Ceryle rudis*), Amur Paradise Flycatcher (*Terpsiphone incei*), Hainan Blue Flycatcher (*Cyornis hainanus*), Tristram's Bunting (*Emberiza tristrami*), Chestnut-eared Bunting (*Emberiza fucata*), Little Bunting (*Emberiza pusilla*), Rustic Bunting (*Emberiza rustica*), Chestnut Bunting (*Emberiza rutila*), Black-faced Bunting (*Emberiza spodocephala*).

Among the wild birds on Hengqin Island, there are eleven species listed on the International Union for Conservation of Nature (IUCN) Red List at the Near Threatened level or higher.

- Two IUCN Red List Endangered (EN) species: Black-faced Spoonbill (*Platalea minor*), Great Knot (*Calidris tenuirostris*).
- Four IUCN Red List Vulnerable (VU) species: Chinese Egret (*Egretta eulophotes*), Common Pochard (*Aythya ferina*), Fairy Pitta (*Pitta nympha*), Rustic Bunting (*Emberiza rustica*).
- Five IUCN Red List Near Threatened (NT) species: Falcated Duck (*Mareca falcate*), Eurasian Curlew (*Numenius arquata*), Grey-tailed Tattler (*Tringa brevipes*), Red-necked Stint (*Calidris ruficollis*), Curlew Sandpiper (*Calidris ferruginea*).

Especially worth mentioning is that Hengqin Island added sixty-four new species of wild birds to its records during the wildlife survey projects from July 2021 to April 2023.

In provincial level, two new species were recorded, namely, Booted Eagle (*Hieraaetus pennatus*) and Black Redstart (*Phonenicurus ochruros*).

Four new species were recorded in Zhuhai: Hartert's Leaf Warbler (*Phylloscopus goodsoni*), Russet Bush Warbler (*Locustella mandelli*), Chestnut Bunting (*Emberiza rutila*), Red-breasted Merganser (*Mergus serrator*).

Fifty-eight new species at Hengqin, including Baikal Teal (*Sibirionetta formosa*), Falcated Duck (*Mareca falcate*), Common Pochard (*Aythya ferina*), Eurasian Spoonbill (*Platalea leucorodia*), Great Bittern (*Botaurus stellaris*), Chinese Egret (*Egretta eulophotes*), Lesser Frigatebird (*Fregata ariel*), Crested Honey-buzzard (*Pernis ptilorhynchus*), Crested Serpent Eagle (*Spilornis cheela*), White-bellied Sea Eagle (*Haliaeetus leucogaster*), Grey-faced Buzzard (*Butastur indicus*), Chinese Sparrowhawk (*Accipiter soloensis*), Slaty-legged Crake (*Rallina eurizonoides*), White-faced Plover (*Charadrius dealbatus*), Ruddy Turnstone (*Arenaria interpres*), Sharp-tailed Sandpiper (*Calidris acuminata*), Long-toed Stint (*Calidris subminuta*), Eurasian Woodcock (*Scolopax rusticola*), Pin-tailed Snipe (*Gallinago stenura*), Oriental Pratincole (*Glareola maldivarum*), Black-tailed Gull (*Larus crassirostris*), Lesser Black-backed Gull (*Larus fuscus*), Little Tern (*Sternula albifrons*), Gull-billed Tern (*Gelochelidon nilotica*), Emerald Dove (*Chalcophaps indica*), Chestnut-winged Cuckoo (*Clamator coromandus*), Himalayan Cuckoo (*Cuculus saturatus*), Eastern Grass Owl (*Tyto longimembris*), Grey Nightjar (*Caprimulgus jotaka*), Blue-tailed Bee-eater (*Merops philippinus*), Common

Hoopoe (*Upupa epops*), Eurasian Wryneck (*Jynx torquilla*), Eurasian Hobby (*Falco subbuteo*), Fairy Pitta (*Pitta nympha*), Grey-chinned Minivet (*Pericrocotus solaris*), Scarlet Minivet (*Pericrocotus speciosus*), Black-naped Monarch (*Hypothymis azurea*), Amur Paradise Flycatcher (*Terpsiphone incei*), Mountain Tailorbird (*Phyllergates cuculatus*), Manchurian Bush Warbler (*Horornis canturians*), Radde's Warbler (*Phylloscopus schwarzi*), Eastern Crowned Warbler (*Phylloscopus coronatus*), Black-browed Reed Warbler (*Acrocephalus bistrigiceps*), Chestnut-collared Yuhina (*Staphida torqueola*), Common Myna (*Acridotheres tristis*), White's Thrush (*Zoothera aurea*), Japanese Thrush (*Turdus cardis*), Eyebrowed Thrush (*Turdus obscurus*), Hainan Blue Flycatcher (*Cyornis hainanus*), Siberian Blue Robin (*Larvivora cyane*), Rufous-tailed Robin (*Larvivora sibilans*), Orange-flanked Bluetail (*Tarsiger cyanurus*), Narcissus Flycatcher (*Ficedula narcissina*), Blue-and-white Flycatcher (*Cyanoptila cyanomelana*), Mugimaki Flycatcher (*Ficedula mugimaki*), Tristram's Bunting (*Emberiza tristrami*), Chestnut-eared Bunting (*Emberiza fucata*), Rustic Bunting (*Emberiza rustica*).

# 横琴野鸟

# The Wild Birds of Hengqin

huā liǎn yā

# 花脸鸭

# Baikal Teal

*Sibirionetta formosa*

**保护等级：**国二 / 无危

**居留类型：**冬候鸟

**体长：**36 ～ 43 厘米

**生活习性：**常与其他鸭类混群活动于河口、湖泊、海湾等湿地生境，或淡水或半咸水；滤食水中浮游动物、软体动物及植物的根茎等。

Protection categories: Ⅱ /LC

Status: Winter visitor

Body length: 36–43 cm

Habitat and Habits: Mix with other ducks in wetlands, like estuaries, lakes, bays, freshwater or brackish water. Filter-feed on plankton, mollusks, and plant roots.

bái méi yā
# 白眉鸭

# Garganey

学 *Spatula querquedula*

**保护等级：**三有 / 无危

**居留类型：**冬候鸟

**体长：**37 ～ 41 厘米

**生活习性：**集群活动于开阔的淡水湖、池塘、河口、沼泽湿地、稻田等。性胆怯且机警，常在有水草隐蔽处活动及觅食。滤食浮游生物、软体动物、昆虫或植物种子。

**Protection categories:** 3V/LC

**Status:** Winter visitor

**Body length:** 37–41cm

**Habitat and Habits:** Clusters in open freshwater like lakes, ponds, river estuaries, marshes, and rice fields. Timid and alert. Seek over among aquatic vegetation to forage. Filter-feed on plankton, mollusks, insects, and plant seeds.

雄鸟♂

雌鸟♀

pí zuǐ yā
# 琵嘴鸭

# Northern Shoveler

㊫ *Spatula clypeata*

**保护等级：**三有 / 无危

**居留类型：**冬候鸟

**体长：**44 ～ 52 厘米

**生活习性：**集大群或与其他鸭类混群活动于河口、湖泊、海湾等湿地生境，相较其他鸭类警惕性不强；滤食水中浮游动物、软体动物及植物的根茎等。

**Protection categories:** 3V/LC

**Status:** Winter visitor

**Body length:** 44–52cm

**Habitat and Habits:** Forms large flocks or associates with other ducks in wetland habitats such as river estuaries, lakes, and bays. Relatively lower vigilance, compared to other ducks. Filter-feed on plankton, mollusks, and the roots and stems of aquatic plants.

luó wén yā

# 罗纹鸭

# Falcated Duck

㊫ *Mareca falcata*

**保护等级：** 三有 / 近危

**居留类型：** 冬候鸟

**体长：** 46 ～ 54 厘米

**生活习性：** 常与其他鸭类混群活动于河口、湖泊、海湾等湿地生境；滤食水中浮游动物、软体动物及植物的根茎等。

Protection categories: 3V /NT

Status: Winter visitor

Body length: 46–54 cm

Habitat and Habits: Often found mingling with other duck species in wetland habitats like river estuaries, lakes, and bays. Filter-feed on plankton, mollusks, and the roots and stems of aquatic plants in the water.

雄鸟 ♂

雌鸟 ♀

chì bǎng yā

# 赤膀鸭

# Gadwall

*Mareca strepera*

**保护等级：** 三有 / 无危

**居留类型：** 冬候鸟

**体长：** 45 ~ 57 厘米

**生活习性：** 常与其他鸭类混群活动于河口、湖泊、海湾等湿地生境，半咸水更常见；滤食水中浮游动物、软体动物及植物的根茎等。

**Protection categories:** 3V / LC

**Status:** Winter visitor

**Body length:** 45–57cm

**Habitat and Habits:** Often seen mingling with other duck species in wetland habitats such as river estuaries, lakes, and bays, with a preference for brackish water. Filter-feed on plankton, mollusks, and the roots and stems of aquatic plants in the water.

chì jǐng yā

# 赤颈鸭

保护等级：三有 / 无危

居留类型：冬候鸟

体长：42 ～ 51 厘米

生活习性：集群生活在海岸边的浅水区或者开阔的浅淡水区，滤食浮游植物、水草等。

# Eurasian Wigeon

学 *Mareca Penelope*

Protection categories：3V / LC

Status：Winter visitor

Body length：42–51cm

Habitat and Habits：Often form large flocks in the shallow water near the coast or in wide-open shallow freshwater; filter-feed on phytoplankton and water plants.

雄鸟 ♂

雌鸟 ♀

bān zuǐ yā

# 斑嘴鸭

**保护等级：** 三有 / 无危

**居留类型：** 冬候鸟

**体长：** 58 ～ 63 厘米

**生活习性：** 常成群栖息于河流、湖泊、水塘、沼泽、水库、滩涂、潟湖等多种生境。常与其他鸭类混群。

# Chinese Spot-billed Duck

学 *Anas zonorhyncha*

**Protection categories:** 3V/LC

**Status:** Winter visitor

**Body length:** 58-63cm

**Habitat and Habits:** Commonly found in groups in various habitats such as rivers, lakes, ponds, marshes, reservoirs, mudflats, and lagoons. Often mix with other duck species.

zhēn wěi yā
# 针尾鸭

# Northern Pintail

学 *Anas acuta*

**保护等级：** 三有 / 无危
**居留类型：** 冬候鸟
**体长：** 51 ～ 76 厘米
**生活习性：** 常成群栖息于沼泽、河流和湖泊等水域，或成对活动于其他鸭类混群中。觅食时头及上身没入水中，臀部朝天。

**Protection categories:** 3V/LC
**Status:** Winter visitor
**Body length:** 51–76cm
**Habitat and Habits:** Often gather in groups in wetlands, rivers, and lakes, or engage in pairs within mixed flocks of other duck species. Head and upper body submerge into water with tail up when foraging.

lǜ chì yā

# 绿翅鸭

**保护等级：** 三有 / 无危

**居留类型：** 冬候鸟

**体长：** 34 ～ 38 厘米

**生活习性：** 常集群或与其他鸭类混群活动于河口、湖泊、水田、海湾等湿地生境。取食植物种子、螺类、昆虫等，也在泥浆中滤食浮游生物。

# Eurasian Teal

㊫ *Anas crecca*

**Protection categories:** 3V/LC

**Status:** Winter visitor

**Body length:** 34–38cm

**Habitat and Habits:** Often form large flocks or mix with other duck species, in rivers, lakes, paddy fields, bays, and other wetland etc.; feed on plant seeds, snails, and insects etc.; also filter-feed on plankton in mud.

雄鸟 ♂

雌鸟 ♀

hóng tóu qián yā

## 红头潜鸭

**保护等级：**三有 / 易危

**居留类型：**冬候鸟

**体长：**41 ～ 50 厘米

**生活习性：**常集群活动于开阔且流速较慢的水域，潜水觅食，常取食水生植物，也会捕食昆虫、螺贝及小鱼。

## Common Pochard

㊫ *Aythya ferina*

**Protection categories:** 3V / VU

**Status:** Winter visitor

**Body length:** 41-50cm

**Habitat and Habits:** Often form large flocks in the wide-open water zone with low flow velocity, diving for food; often feed on water plants, insects, snails, shellfish and fish.

bān bèi qián yā

# 斑背潜鸭

**保护等级：** 三有 / 无危

**居留类型：** 冬候鸟

**体长：** 42 ~ 49 厘米

**生活习性：** 多在沿海水域或河口活动，有时光顾淡水湖泊。合群且以群栖居。

# Greater Scaup

㊫ *Aythya marila*

**Protection categories:** 3V / LC

**Status:** Winter visitor

**Body length:** 42–49cm

**Habitat and Habits:** Often found in coastal waters or river estuaries, and occasionally visit freshwater lakes. Inhabit in groups.

fèng tóu qián yā

# 凤头潜鸭

保护等级：三有 / 无危

居留类型：冬候鸟

体长：34 ～ 49 厘米

生活习性：集大群栖息深水处随波漂浮，潜水觅食螺贝类，也吃水生植物种子、嫩叶及水生昆虫。

# Tufted Duck

㊫ *Aythya fuligula*

Protection categories：3V / LC

Status：Winter visitor

Body length：34–49cm

Habitat and Habits：Often form large flocks floating in the deep water zone with the wave; diving for snails and shellfish, and also feed on seeds of water plant, young leaves and aquatic insects.

hóng xiōng qiū shā yā

# 红胸秋沙鸭

# Red-breasted Merganser

㊫ *Mergus serrator*

**保护等级：** 三有 / 无危

**居留类型：** 冬候鸟

**体长：** 52 ~ 60 厘米

**生活习性：** 偏好沿海咸水生境，集小群活动，潜水捕鱼为食，会将头先伸入水下寻找食物再潜水捕捉。

**Protection categories:** 3V / LC

**Status:** Winter visitor

**Body length:** 52–60cm

**Habitat and Habits:** Prefer coastal saline habitats and gather in small groups. Dive for fish, and search for food with heads submerged before diving to catch it.

zhōng huá zhè gū

# 中华鹧鸪

## Chinese Francolin

㊫ *Francolinus pintadeanus*

**保护等级：** 三有 / 无危

**居留类型：** 留鸟

**体长：** 29 ～ 35 厘米

**生活习性：** 栖于低地至海拔 1600 米的干燥林地、草地及次生灌丛。

Protection categories：3V/LC

Status：Resident bird

Body length：29–35cm

Habitat and Habits：Inhabit dry forests, grasslands, and secondary shrublands, ranging from lowlands to elevations of 1600 meters.

xiǎo pì tī
# 小䴙䴘

保护等级：三有 / 无危
居留类型：留鸟
体长：23 ～ 29 厘米
生活习性：一般活动于湖泊、水塘、沼泽等淡水区域，潜水觅食水生动物。起飞须助跑如“轻功水上漂”。育雏期间发现危险时会将幼鸟背在背上。

# Little Grebe

㊫ *Tachybaptus ruficollis*

Protection categories：3V/LC
Status：Resident bird
Body length：23–29cm
Habitat and Habits：Often inhabit lakes，ponds，marshes，and other freshwater，diving for aquatic animals；need to run for taking off，looking like the Chinese Kung Fu flight on water；the parents will carry their young on their back at the first sign of danger.

bái pí lù

# 白琵鹭

# Eurasian Spoonbill

㊫ *Platalea leucorodia*

**保护等级：**国二 / 无危

**居留类型：**冬候鸟

**体长：**80 ～ 95 厘米

**生活习性：**常集群或混群活动于淡水及半咸水湿地，如鱼虾塘、浅水湖泊等，觅食时琵琶状的嘴在水中左右横扫。主要捕食鱼类及软体动物等。

Protection categories: II/LC

Status: Winter visitor

Body length: 80–95cm

Habitat and Habits: Cluster or mix in freshwater and brackish wetlands, such as fish and shrimp ponds, shallow lakes, etc. Sweeping their spoon-shaped bills while foraging for fish and mollusks.

hēi liǎn pí lù

# 黑脸琵鹭

# Black-faced Spoonbill

㊫ *Platalea minor*

**保护等级：** 国一 / 濒危

**居留类型：** 冬候鸟

**体长：** 60 ~ 79 厘米

**生活习性：** 常集群活动于沿海滩涂、鱼虾塘、湖泊等，觅食时将琵琶状嘴伸入水中左右横扫，捕食鱼虾、无脊椎动物等。

**Protection categories:** I/EN

**Status:** Winter visitor

**Body length:** 60–79cm

**Habitat and Habits:** Often form flocks in the coastal beaches, fish and shrimp ponds, and lakes, etc.; when foraging, stretch the spoon-shaped bill into water and sweeps around; prey on fish, shrimp, and invertebrates, etc.

繁殖羽

Breeding plumage

国家一级保护野生动物：黑脸琵鹭

The national first-class protected wild animal: Black-faced Spoonbill

dà má jiān
# 大麻鳽

## Great Bittern

㊫ *Botaurus stellaris*

**保护等级：** 三有 / 无危

**居留类型：** 冬候鸟

**体长：** 64 ～ 78 厘米

**生活习性：** 常单独活动于溪流、湖泊等淡水及咸淡水湿地里的芦苇上，伪装能力强，受惊时头部竖起，一动不动。主要捕食湿地里的两栖类、爬行类、鱼类、软体动物、昆虫等。

**Protection categories：** 3V / LC

**Status：** Winter visitor

**Body length：** 64–78cm

**Habitat and Habits：** Perch alone on stream, lakes, freshwater and brackish wetlands, with amazing camouflage skills. Remain motionless with raised heads when alarmed. Feed on amphibians, fish, mollusks, and insects found in wetland habitats.

huáng bān wěi jiān
# 黄斑苇鳽

# Yellow Bittern
㊫ *Ixobrychus sinensis*

**保护等级：**三有 / 无危

**居留类型：**留鸟

**体长：**30 ～ 40 厘米

**生活习性：**常单独活动于芦苇丛、荷花池、稻田等。保护色极好，常常双脚分别站立于植物秆上，伺机迅速捕食鱼类、两栖类、昆虫等。捕食时颈部可瞬间伸长 2 倍。

**Protection categories:** 3V/LC

**Status:** Resident bird

**Body length:** 30–40cm

**Habitat and Habits:** Often inhabit alone in reeds, lotus ponds, and paddy fields, etc.; Amazing plumage camouflage, and straddle reeds, waiting for an opportunity to quickly prey on fish, amphibians, and insects, etc.; when foraging, the neck can be instantly extended twice longer.

lì wěi jiān

# 栗苇鳽

# Cinnamon Bittern

㊫ *Ixobrychus cinnamomeus*

**保护等级：** 三有 / 无危

**居留类型：** 留鸟

**体长：** 40 ～ 41 厘米

**生活习性：** 常单独活动，隐匿于水边芦苇丛或灌丛中。捕食鱼类、两栖类和昆虫等小动物。

**Protection categories：** 3V / LC

**Status：** Resident bird

**Body length：** 40–41 cm

**Habitat and Habits：** Hide solitarily in reeds and bushes near water. Feed on fish, amphibians, insects and other small creatures.

yè lù
# 夜鹭

# Black-crowned Night Heron

学 *Nycticorax nycticorax*

**保护等级：** 三有 / 无危

**居留类型：** 留鸟

**体长：** 58 ～ 65 厘米

**生活习性：** 活动于人工湿地、红树林等各种湿地环境，昼夜均可捕食，常在水边等待，迅速捕食鱼虾、两栖类等。集群在树上筑巢繁殖。

**Protection categories:** 3V/LC

**Status:** Resident bird

**Body length:** 58–65cm

**Habitat and Habits:** Inhabit artificial wetland, mangrove and any other wetlands; forage in the morning and night; often wait near the water and quickly prey on fish, shrimps, and amphibians, etc.; often form flocks to build nests and breed on trees.

lǜ lù
# 绿鹭

# Striated Heron

㊫ *Butorides striata*

**保护等级：** 三有 / 无危

**居留类型：** 留鸟

**体长：** 35 ～ 48 厘米

**生活习性：** 常单独活动于河流、水库、湖泊或红树林，在水边长时间静候，捕食鱼虾、两栖类等。

**Protection categories:** 3V/LC

**Status:** Resident bird

**Body length:** 35–48cm

**Habitat and Habits:** Often inhabit rivers, reservoirs, lakes or mangroves; stand and wait near water for long time, to prey on fish, shrimps, and amphibians, etc.

chí lù
# 池鹭

# Chinese Pond Heron

㊫ *Ardeola bacchus*

**保护等级：** 三有 / 无危

**居留类型：** 留鸟

**体长：** 42 ~ 52 厘米

**生活习性：** 常单独活动于各种湿地，捕食鱼类、两栖类、昆虫等小型动物。

Protection categories: 3V/LC

Status: Resident bird

Body length: 42–52cm

Habitat and Habits: Often inhabit alone in all kinds of wetland; prey on fish, amphibians, insects and other small animals.

繁殖羽 br.

非繁殖羽 non-br.

niú bèi lù

# 牛背鹭

# Eastern Cattle Egret

㊫ *Bubulcus coromandus*

**保护等级：** 三有

**居留类型：** 留鸟

**体长：** 46 ～ 53 厘米

**生活习性：** 常活动于农田、牧场，或低山的湿地，主要在湿地和草地觅食昆虫及两爬类，常与牛相伴，站立于牛背上，捕食被牛惊飞的昆虫。

**Protection categories:** 3V

**Status:** Resident bird

**Body length:** 46-53cm

**Habitat and Habits:** Often inhabit fields, farms, and wetlands in the low hilly areas; prey on insects, amphibians and reptilians; often stay with cattle and catch insects disturbed by the cattle.

繁殖羽 br.

非繁殖羽 non-br.

cāng lù
# 苍鹭

# Grey Heron

㊫ *Ardea cinerea*

**保护等级：**三有 / 无危

**居留类型：**留鸟

**体长：**92～99 厘米

**生活习性：**活动于河流、湖泊、红树林滩涂等多种湿地，常在浅水中长时间站立，伺机捕食鱼类、两栖类和各种小型动物，俗称“老等”。

**Protection categories:** 3V/LC

**Status:** Resident bird

**Body length:** 92–99cm

**Habitat and Habits:** Inhabit rivers, lakes, beaches of mangrove mudflat and other wetland habitats; often stand in the shallow water for a long time and wait for an opportunity to prey on fish, amphibians, and other small animals.

cǎo lù
# 草鹭

# Purple Heron

Ⓢ *Ardea purpurea*

**保护等级：**三有 / 无危

**居留类型：**留鸟

**体长：**84 ～ 97 厘米

**生活习性：**喜稻田、芦苇地、湖泊及溪流。性孤僻，常单独在有芦苇的浅水中，低歪着头伺机捕鱼及其他食物。

**Protection categories:** 3V/LC

**Status:** Resident bird

**Body length:** 84–97cm

**Habitat and Habits:** Inhabit paddy field, reed, lakes and streams; Stay solitarily in shallow water with reeds and catch fish and other creatures.

dà bái lù
# 大白鹭

# Great Egret

㊫ *Ardea alba*

**保护等级：**三有 / 无危

**居留类型：**留鸟

**体长：**90 ～ 98 厘米

**生活习性：**活动于河流、湖泊、红树林滩涂等多种湿地，常静立水中或缓慢行走伺机捕食鱼类、两栖类和各种小型动物。

**Protection categories:** 3V/LC

**Status:** Resident bird

**Body length:** 90–98cm

**Habitat and Habits:** Inhabit rivers, lakes, mangroves, and other wetland habitats; often stand in water quietly, or walk slowly to forage fish, amphibians, and other small animals.

zhōng bái lù
# 中白鹭

# Intermediate Egret

*Ardea intermedia*

**保护等级：** 三有 / 无危

**居留类型：** 留鸟

**体长：** 62 ~ 70 厘米

**生活习性：** 常单独或集群活动于湿地草丛中，偏好淡水环境。主要取食湿地里的两栖类、爬行类、鱼类、软体动物等。

**Protection categories:** 3V / LC

**Status:** Resident bird

**Body length:** 62–70cm

**Habitat and Habits:** Often forage alone or in groups within wetland vegetation, preferring freshwater environments. Their diet primarily consists of amphibians, reptiles, fish, and mollusks.

bái lù
# 白鹭

# Little Egret

㊫ *Egretta garzetta*

**保护等级：** 三有 / 无危

**居留类型：** 留鸟

**体长：** 55 ～ 68 厘米

**生活习性：** 常集群活动于红树林、滩涂、池塘等各种湿地环境，觅食时常以脚搅动水底，把鱼虾类等猎物惊吓出来迅速捕食。集群繁殖，筑巢于树上。

**Protection categories：** 3V/LC

**Status：** Resident bird

**Body length：** 55-68cm

**Habitat and Habits：** Often form flocks in mangroves, beaches, ponds and all kinds of wetlands; stir the water at the bottom with feet when foraging, to startle fish and shrimps and other prey out and quickly catch them; breed together and build the nest on trees.

国家二级保护野生动物：岩鹭
The national second-class protected wild animal: Pacific Reef Heron

yán lù
# 岩鹭

# Pacific Reef Heron

㊫ *Egretta sacra*

**保护等级：** 国二 / 无危

**居留类型：** 留鸟

**体长：** 58 ～ 66 厘米

**生活习性：** 常单独活动于海岛、海岸带的岩石上，或在岩石上鬼祟前行，或主动飞行空中捕食鱼虾。

Protection categories：II/LC

Status：Resident bird

Body length：58–66cm

Habitat and Habits：Often stay alone on the rocks nearby the coasts；often forage quietly or move slowly on the rocks；mostly prey on small fish and crabs，sometime prey on insects and mollusks.

huáng zuǐ bái lù

# 黄嘴白鹭

**保护等级：** 国一 / 易危

**居留类型：** 过境鸟

**体长：** 65 ~ 68 厘米

**生活习性：** 常觅食于海湾和河流出海口环境的湿地，偏好海水环境。主食鱼类、两栖类及水中无脊椎动物。营巢于无人海岛的峭壁或树丛。

# Chinese Egret

㊫ *Egretta eulophotes*

**Protection categories:** I/VU

**Status:** Passing visitor

**Body length:** 65–68cm

**Habitat and Habits:** Often forage in the wetlands of bays and coastal estuaries, and prefer seawater environment; mainly prey on fish, amphibians and aquatic invertebrates; make nests on the bush of the rocky cliffs of uninhabited islands.

国家一级保护野生动物：黄嘴白鹭

The national first-class protected wild animal: Chinese Egret

国家二级保护野生动物：白斑军舰鸟

The national second-class protected wild animal: Lesser Frigatebird

bái bān jūn jiàn niǎo

# 白斑军舰鸟

# Lesser Frigatebird

㊫ *Fregata ariel*

**保护等级：** 国二 / 无危

**居留类型：** 夏候鸟

**体长：** 68 ～ 81 厘米

**生活习性：** 常单独或集小群在海面上空飞翔，善于飞行，偶尔沿着海岸线飞翔。头部插入水中捕鱼，或者掠夺其他鸟类捕捉到的鱼类。

Protection categories： II/LC

Status： Summer visitor

Body length： 68–81cm

Habitat and Habits： Fly solo or in small groups above the sea, displaying skilled flight. Dip heads into water to catch fish or scavenge from other birds' catches.

pǔ tōng lú cí
# 普通鸬鹚

# Great Cormorant

㊫ *Phalacrocorax carbo*

**保护等级：**三有 / 无危

**居留类型：**冬候鸟

**体长：**77 ～ 94 厘米

**生活习性：**常集群活动于环境相对安全的孤岛石头上或树上，善于潜水。以鱼类为食，潜入水中捕捉。

**Protection categories：** 3V / LC

**Status：** Winter visitor

**Body length：** 77–94cm

**Habitat and Habits：** Gather in clusters on isolated rocks or trees where is relatively safe. Excellent in diving. Dive for and feed mainly on fish.

è
# 鹗

保护等级：国二 / 无危
居留类型：留鸟
体长：56 ～ 62 厘米
生活习性：活动于海岸、水库、湖泊等鱼类丰富的湿地，捕鱼之鹰，俯冲入水抓活鱼。

# Western Osprey

㊫ *Pandion haliaetus*

Protection categories：II / LC
Status：Resident bird
Body length：56–62 cm
Habitat and Habits：Inhabit coast, reservoirs, lakes and other wetlands with abundant fish resources; catch fish by spotting them from the air, diving and plunging into the water to catch fish.

hēi chì yuān
# 黑翅鸢

# Black-winged Kite

㊫ *Elanus caeruleus*

**保护等级：** 国二 / 无危
**居留类型：** 留鸟
**体长：** 31 ～ 37 厘米
**生活习性：** 单独活动于水稻田、荒地等开阔环境，善悬停寻找猎物，捕食鼠类、蜥蜴、小型鸟类、昆虫。

**Protection categories:** II/LC
**Status:** Resident bird
**Body length:** 31–37cm
**Habitat and Habits:** Inhabit alone in paddy fields, uncultivated land and open country; good at hovering when in search for prey; prey on rats and mice, lizards, small birds, and insects.

fèng tóu fēng yīng

# 凤头蜂鹰

**保护等级：** 国二 / 无危

**居留类型：** 过境鸟

**体长：** 57～61 厘米

**生活习性：** 秋季迁徙季节常集大群或混群迁徙，多见于半山腰或水岸交接，有上升气流的环境。主要捕食蜂类等昆虫。

# Crested Honey-buzzard

㊫ *Pernis ptilorhynchus*

**Protection categories:** II/LC

**Status:** Passing visitor

**Body length:** 57–61 cm

**Habitat and Habits:** Migrate large flocks or mixed groups during autumn, often seen at mid-altitudes or where mountains meet water. Mainly feed on insects, including bees.

shé diāo

# 蛇雕

# Crested Serpent Eagle

*Spilornis cheela*

**保护等级：**国二 / 无危

**居留类型：**留鸟

**体长：**65 ~ 74 厘米

**生活习性：**活动于较原始的阔叶林区，也可到山区果园等林地。捕食爬行类、鼠类及鸟类。

Protection categories: II/LC

Status: Resident bird

Body length: 65–74cm

Habitat and Habits: Inhabit primitive broad-leaved forest, fruit-bearing forest and other woodland; prey on reptiles, rats, mice and birds.

国家二级保护野生动物：蛇雕

The national second-class protected wild animal: Crested Serpent Eagle

xuē sǔn diāo

# 靴隼雕

# Booted Eagle

学 *Hieraaetus pennatus*

**保护等级：**国二 / 无危

**居留类型：**过境鸟

**体长：**42 ～ 51 厘米

**生活习性：**常单独或成对于林缘的空中来回飞行或盘旋寻找猎物。捕食小型兽类、鼠类及鸟类。

Protection categories：II/LC

Status：Passing visitor

Body length：42–51cm

Habitat and Habits：Alone or in pairs，fly around or hover in the forest edge to find preys；prey on small mammals and birds.

chì fù yīng

# 赤腹鹰

# Chinese Sparrowhawk

㊫ *Accipiter soloensis*

**保护等级：** 国二 / 无危

**居留类型：** 过境鸟

**体长：** 25 ～ 35 厘米

**生活习性：** 秋季迁徙季节常集大群或混群迁徙，夏季单独或成对活动，繁殖期领域性强，常驱赶其他鸟类。主要捕食昆虫及其他小型动物。

**Protection categories:** II/LC

**Status:** Passing visitor

**Body length:** 25–35cm

**Habitat and Habits:** Migrate in large flocks or mixed groups during autumn. Solitary or in pairs in summer. Exhibit territorial behavior during breeding season by driving away other birds. Diet consists mainly of insects and other small animals.

fèng tóu yīng
# 凤头鹰

# Crested Goshawk

㊫ *Accipiter trivirgatus*

**保护等级：** 国二 / 无危

**居留类型：** 留鸟

**体长：** 40 ～ 48 厘米

**生活习性：** 活动于低海拔的林地。繁殖期飞行可见其白色蓬松的尾下覆羽如“尿不湿”，有快速抖动双翼的特点。捕食鼠类、两栖类、鸟类等。

**Protection categories:** II/LC

**Status:** Resident bird

**Body length:** 40–48cm

**Habitat and Habits:** Inhabit woodlands with low altitude; during the breeding season, white fluffy under-tail coverts could be seen when flying; flap the wings quickly; prey on rats, mice, amphibians and birds, etc.

国家二级保护野生动物：凤头鹰
The national second-class protected wild animal: Crested Goshawk

sōng què yīng
# 松雀鹰

# Besra

㊫ *Accipiter virgatus*

**保护等级：** 国二 / 无危

**居留类型：** 留鸟

**体长：** 25～36 厘米

**生活习性：** 秋季迁徙季节常集大群或混群迁徙，夏季繁殖时主要活动于林相较好的林地，领域性强。主要以小型鸟类，小型两栖爬行类和昆虫为食。

**Protection categories:** II/LC

**Status:** Resident bird

**Body length:** 25–36cm

**Habitat and Habits:** During autumn migration, migrate in large flocks. During summer breeding season, inhabit forest and become territorial, mainly prey on small birds, amphibians, reptiles, and insects.

què yào
# 鹊鹞

# Pied Harrier

㊫ *Circus melanoleucos*

**保护等级：** 国二 / 无危

**居留类型：** 冬候鸟

**体长：** 43～50 厘米

**生活习性：** 常单独活动于水稻田及其他水草茂盛的湿地生态环境，飞行高度不高，几乎贴地飞行。主要捕食小型动物，如小型鸟类。

**Protection categories:** II/LC

**Status:** Winter visitor

**Body length:** 43–50cm

**Habitat and Habits:** Solitary foraging in rice fields and wetlands with dense aquatic vegetation. Fly low near the ground. Mainly prey on small animals, like small birds.

bái fù yào
# 白腹鹞

# Eastern Marsh Harrier

*Circus spilonotus*

**保护等级：** 国二 / 无危

**居留类型：** 冬候鸟

**体长：** 48 ～ 58 厘米

**生活习性：** 常栖息于原野、沼泽及农田等开阔生境，经常低飞，翅膀呈浅 V 形，以鼠类和小型鸟类为食。

**Protection categories:** II/LC

**Status:** Winter visitor

**Body length:** 48–58cm

**Habitat and Habits:** Commonly inhabit wide-open habitats such as grasslands, marshes, and farmland. Often fly low with shallow V-shaped wings; primarily feed on rats, mice and small birds.

huī liǎn kuáng yīng

# 灰脸鵟鹰

# Grey-faced Buzzard

㊫ *Butastur indicus*

**保护等级：** 国二 / 无危

**居留类型：** 过境鸟

**体长：** 39 ～ 48 厘米

**生活习性：** 单独或成对活动，偏好林地边缘，迁徙期才集大群。捕食两爬类、鼠类及大型昆虫。

Protection categories: II/LC

Status: Passing visitor

Body length: 39–48cm

Habitat and Habits: Solitary or paired, prefer forest edges, gather in large groups during migration. Prey on amphibians, reptiles, rats, mice and large insects.

bái fù hǎi diāo
# 白腹海雕

# White-bellied Sea Eagle

㊫ *Haliaeetus leucogaster*

**保护等级：** 国一 / 无危

**居留类型：** 留鸟

**体长：** 70 ～ 85 厘米

**生活习性：** 活动于海岛、海岸及河口，有时也在海岸附近的水库和丘陵出现，常在海域以双爪入水捕鱼及其他海洋小动物。

Protection categories: I/LC

Status: Resident bird

Body length: 70–85cm

Habitat and Habits: Inhabit islands, coasts, and estuaries, sometimes in reservoirs and hills near the coasts; often prey over the sea and catch fish and other small marine animals with strong claws.

国家一级保护野生动物：白腹海雕

The national first-class protected wild animal: White-bellied Sea Eagle

国家二级保护野生动物：黑鸢
The national second-class protected wild animal: Black Kite

hēi yuān
# 黑鸢

# Black Kite

㊫ *Milvus migrans*

**保护等级：** 国二 / 无危

**居留类型：** 留鸟

**体长：** 54 ～ 66 厘米

**生活习性：** 珠海最常见的猛禽，俗称"麻鹰"。集群活动于较开阔的湿地、林地上空，甚至于城市楼宇间穿梭，寻觅鱼类、两栖类、鼠类为食，亦食腐肉。

**Protection categories：** II / LC

**Status：** Resident bird

**Body length：** 54–66cm

**Habitat and Habits：** Most common seen raptors in Zhuhai; form flocks in open wetlands and woodland, and also in urban city; prey on fish, amphibians, rats and mice; and also carrion.

pǔ tōng kuáng
# 普通鵟

# Eastern Buzzard

㊫ *Buteo japonicus*

**保护等级：** 国二 / 无危

**居留类型：** 冬候鸟

**体长：** 42 ～ 54 厘米

**生活习性：** 对环境不挑剔，可见于林地上空，也可见于草原、湿地等旷野地区，觅食时有悬停行为，主食鼠类。

Protection categories: II/LC

Status: Winter visitor

Body length: 42–54cm

Habitat and Habits: Live in a range of habitats; can be seen in the woodland, grassland, wetland, and other open areas; hover when foraging; mainly prey on rats and mice.

国家二级保护野生动物：普通鵟

The national second-class protected wild animal: Eastern Buzzard

pǔ tōng yāng jī

# 普通秧鸡

# Brown-cheeked Rail

㊫ *Rallus indicus*

**保护等级：** 三有 / 无危

**居留类型：** 冬候鸟

**体长：** 23 ～ 39 厘米

**生活习性：** 常单独活动于湿地隐秘的位置，喜水边茂密植被，甚少飞行。主要以水边生活的生物为食，如软体动物及昆虫等。

Protection categories: 3V/LC

Status: Winter visitor

Body length: 23–39cm

Habitat and Habits: Solitary, favor hiding among the dense vegetation of wetland habitats, rarely fly. Primarily feed on aquatic creatures such as mollusks and insects.

huī xiōng yāng jī

# 灰胸秧鸡

# Slaty-breasted Rail

㊫ *Lewinia striatia*

**保护等级：** 三有 / 无危

**居留类型：** 留鸟

**体长：** 25 ～ 30 厘米

**生活习性：** 单独活动于湿地，包括沼泽、芦苇、稻田、河岸、红树林、虾塘等，多见于半咸水湿地。主要以水边生活的软体动物等生物为食。

**Protection categories:** 3V/LC

**Status:** Resident bird

**Body length:** 25–30cm

**Habitat and Habits:** Solitary, in wetlands like marshes, reed beds, rice fields, riverbanks, mangroves, and brackish water. Feed on mollusks and other aquatic organisms.

hēi shuǐ jī
# 黑水鸡

# Common Moorhen

㊫ *Gallinula chloropus*

**保护等级：** 三有 / 无危

**居留类型：** 留鸟

**体长：** 30 ～ 38 厘米

**生活习性：** 常于浮水植物中（如荷花、睡莲、水葫芦等）活动，善游泳，游泳时头部规律性前伸。主要以水生植物、昆虫、蝌蚪为食。

**Protection categories:** 3V/LC

**Status:** Resident bird

**Body length:** 30–38cm

**Habitat and Habits:** Often inhabit wetlands with floating plants, such as lotus, water lily, water hyacinth, etc.; good at swimming; the head is regularly extended forward when swimming; mainly feed on aquatic plants, insects, and tadpoles.

bái gǔ dǐng

# 白骨顶

# Eurasian Coot

㊫ *Fulica atra*

**保护等级：**三有 / 无危

**居留类型：**冬候鸟

**体长：**36 ～ 39 厘米

**生活习性：**栖息于浮水植物中，或者开阔的浅水区域，善游泳也善潜水。以水生植物、昆虫为食。

**Protection categories：** 3V/LC

**Status：** Winter visitor

**Body length：** 36–39cm

**Habitat and Habits：** Inhabit wetlands with floating plants or open shallow water areas; good at swimming and diving; feed on aquatic plants and insects.

bái xiōng kǔ è niǎo

# 白胸苦恶鸟

# White-breasted Waterhen

*Amaurornis phoenicurus*

**保护等级：** 三有 / 无危

**居留类型：** 留鸟

**体长：** 28 ～ 33 厘米

**生活习性：** 性隐秘，躲藏在草丛中，常叫。遇到危险时迅速半飞半跑进入灌丛中。以植物嫩芽、软体动物、昆虫、螺贝为食。

**Protection categories:** 3V/LC

**Status:** Resident bird

**Body length:** 28–33cm

**Habitat and Habits:** Shy, often hide in bushes; often calls; quickly run and fly into bushes when in danger; feed on plant buds, mollusks and insects.

hóng xiōng tián jī

## 红胸田鸡

**保护等级：** 三有 / 无危

**居留类型：** 冬候鸟

**体长：** 19 ～ 23 厘米

**生活习性：** 常单独活动于河道、水稻田、芦苇滩及草丛等淡水湿地。主要以软体动物及各种水生生物为食。

## Ruddy-breasted Crake

㊫ *Zapornia fusca*

**Protection categories：** 3V/LC

**Status：** Winter visitor

**Body length：** 19–23cm

**Habitat and Habits：** Solitary in freshwater wetlands such as rivers, rice fields, reed beds, and grassy areas. Feed primarily on mollusks and various aquatic creatures.

bái hóu bān yāng jī

## 白喉斑秧鸡

**保护等级：** 三有 / 无危

**居留类型：** 留鸟

**体长：** 21 ～ 28 厘米

**生活习性：** 常单独活动于森林中的水道边，夜行性，善鸣叫，白天常躲在密林中休息。主要以林层地上的动物为食，如各类环节动物。

## Slaty-legged Crake

㊫ *Rallina eurizonoides*

**Protection categories：** 3V/LC

**Status：** Resident bird

**Body length：** 21–28cm

**Habitat and Habits：** Solitary in forest areas near waterways. Nocturnal and hide in dense forest grounds during day time. Good at singing. Feed primarily on insects and other small creatures under the forest grounds.

hēi chì cháng jiǎo yù

# 黑翅长脚鹬

# Black-winged Stilt

㊫ *Himantopus himantopus*

**保护等级：**三有 / 无危

**居留类型：**冬候鸟

**体长：**35 ～ 40 厘米

**生活习性：**集群觅食于淡水浅水区，也可见于河口偏淡水区域，喜水稻田。捕食水生昆虫、甲壳类、小鱼及蝌蚪。繁殖期亲鸟有非常激烈的保护巢或幼鸟的行为。

Protection categories：3V / LC

Status：Winter visitor

Body length：35–40cm

Habitat and Habits：Form flocks in shallow freshwater areas or partial freshwater areas in estuaries, and like feeding on paddy fields; prey on aquatic insects, small fish, and tadpoles; the parents have the fierce behaviors of protecting nests and juveniles when breeding.

黑翅长脚鹬 Black-winged Stilt

fǎn zuǐ yù
# 反嘴鹬

# Pied Avocet

*Recurvirostra avosetta*

**保护等级：**三有 / 无危
**居留类型：**冬候鸟
**体长：**42 ~ 45 厘米
**生活习性：**冬季集大群活动于湖泊、沼泽、海岸浅滩中，擅长游泳，并会在水中倒立，觅食时用喙左右扫动。

**Protection categories:** 3V / LC
**Status:** Winter visitor
**Body length:** 42–45cm
**Habitat and Habits:** During winter, large flocks in lakes, marshes, shallow coastal areas. Skilled swimmers, up-ending like a duck to reach food below the surface, also can feed by sweeping bills from side to side in water.

huī tóu mài jī

# 灰头麦鸡

# Grey-headed Lapwing

学 *Vanellus cinereus*

**保护等级：** 三有 / 无危

**居留类型：** 冬候鸟

**体长：** 34 ～ 37 厘米

**生活习性：** 常集群或混群活动于各类湿地，以淡水湿地为主，尤其是浅水的低矮草地。主要觅食沼泽地中的软体动物及无脊椎动物等。

**Protection categories:** 3V/LC

**Status:** Winter visitor

**Body length:** 34–37cm

**Habitat and Habits:** Groups or mixes in diverse wetlands, primarily freshwater habitats, especially in shallow water with low grasslands. Feeds mainly on wetland invertebrates like mollusks and annelids.

huī héng
# 灰鸻

# Grey Plover

㊫ *Pluvialis squatarola*

**保护等级：** 三有 / 无危

**居留类型：** 冬候鸟

**体长：** 27 ～ 31 厘米

**生活习性：** 常集小群活动于海边及河口处的泥滩，也与其他鸻鹬类混群觅食。主要取食泥滩中各类小虾及泥滩生活的水生生物。

**Protection categories:** 3V/LC

**Status:** Winter visitor

**Body length:** 27–31cm

**Habitat and Habits:** Gathers in small groups on coastal mudflats and river estuaries, often forages with other sandpipers. Feeds mainly on small shrimps and other aquatic organisms in the mudflats.

jīn héng
# 金鸻

# Pacific Golden Plover

㊫ *Pluvialis fulva*

**保护等级：** 三有 / 无危

**居留类型：** 冬候鸟

**体长：** 23 ～ 26 厘米

**生活习性：** 栖息于河岸附近的农田、水塘、沼泽及空旷原野上，以植物种子、嫩芽、软体动物、甲壳类和昆虫为食。

**Protection categories:** 3V/LC

**Status:** Winter visitor

**Body length:** 23–26cm

**Habitat and Habits:** Inhabit farmlands, ponds, swamps and open fields; feed on plant seeds, buds, mollusks, crustaceans and insects.

měng gǔ shā héng

# 蒙古沙鸻

# Lesser Sand Plover

*Charadrius mongolus*

**保护等级：** 三有 / 无危

**居留类型：** 冬候鸟

**体长：** 18 ～ 21 厘米

**生活习性：** 常集大群或混群活动于河口、海边泥沙滩、盐田等，也偶见于内陆河湿地，行走速度快，休息时常单足站立。取食滩涂各类小型水生生物。

**Protection categories:** 3V/LC

**Status:** Winter visitor

**Body length:** 18–21cm

**Habitat and Habits:** Forms flocks or mix with other groups on estuaries, beaches, and salt marshes, and also found in inland wetlands. Runs fast. Stands on one leg while resting. Feeds on small aquatic organisms in tidal flats.

tiě zuǐ shā héng
# 铁嘴沙鸻

# Greater Sand Plover

㊫ *Charadrius leschenaultii*

**保护等级：** 三有 / 无危

**居留类型：** 冬候鸟

**体长：** 22 ～ 25 厘米

**生活习性：** 常小群或分散活动于河口、海边泥沙滩等，行动相对较快。取食滩涂各类小型水生生物。

**Protection categories：** 3V / LC

**Status：** Winter visitor

**Body length：** 22–25cm

**Habitat and Habits：** Often found in small group or individually along estuaries and sandy beaches. Moves quickly. Feeds on various small aquatic organisms in tidal flats.

jīn kuàng héng

# 金眶鸻

# Little Ringed Plover

*Charadrius dubius*

**保护等级：** 三有 / 无危

**居留类型：** 留鸟

**体长：** 14～17 厘米

**生活习性：** 偏好淡水环境，常活动于鱼塘、水田等湿地。觅食时常抖动脚把隐藏的水生生物或环节动物惊吓出来啄食。

Protection categories：3V/LC

Status：Resident bird

Body length：14–17cm

Habitat and Habits：Prefer freshwater，and often inhabit fishponds，paddy fields and other wetlands；scare hidden aquatic creatures or annelids out and peck at them by shacking the feet when foraging.

bái liǎn héng
# 白脸鸻

# White-faced Plover

㊫ *Charadrius dealbatus*

**保护等级：**无

**居留类型：**夏候鸟

**体长：**15 ~ 17 厘米

**生活习性：**除繁殖期外常集大群活动，主要在沙滩上活动，善行走，“走—停—吃”是其典型行为。啄食泥滩上的昆虫、节肢动物及软体动物。

Protection categories: DD

Status: Summer visitor

Body length: 15-17cm

Habitat and Habits: Often form large flocks excepting the breeding season; mainly inhabit beaches; good at walking; their classical behavior is “walk-stop-eat”; peak at insects, arthropods, and mollusk in the mudbanks.

huán jǐng héng

# 环颈鸻

# Kentish Plover

㊫ *Charadrius alexandrinus*

**保护等级：**三有 / 无危

**居留类型：**冬候鸟

**体长：**15 ～ 17 厘米

**生活习性：**除繁殖期外常集大群活动，主要在沙滩或泥滩上活动，善行走，“走—停—吃”是其典型行为。啄食泥滩上的昆虫、节肢动物及软体动物。

Protection categories: 3V/LC

Status: Winter visitor

Body length: 15–17cm

Habitat and Habits: Often form large flocks excepting the breeding season; mainly inhabit beaches or mudbanks; good at walking; their classical behavior is “walk-stop-eat”; peak at insects, arthropods, and mollusk in the mudbanks.

cǎi yù
# 彩鹬

# Greater Painted Snipe

㊫ *Rostratula benghalensis*

**保护等级：**三有 / 无危

**居留类型：**留鸟

**体长：**23 ～ 28 厘米

**生活习性：**常集小群活动于各类湿地，尤其是水稻田、低矮草地等，性隐秘且警惕性强。食谱较广，包括小型水生生物及植物嫩芽等。

**Protection categories:** 3V/LC

**Status:** Resident bird

**Body length:** 23–28cm

**Habitat and Habits:** Inhabit in small group in wetlands, especially rice paddies and low grasslands. Secretive and vigilant. Feed on a variety food, including small aquatic organisms and tender plant shoots.

雌鸟 ♀

雄鸟 ♂

zhōng sháo yù
# 中杓鹬

**保护等级：**三有 / 无危

**居留类型：**过境鸟

**体长：**40 ~ 46 厘米

**生活习性：**活动于沿海潮间带的礁石、泥滩、沙滩等湿地及岸边草地。喜吃蟹类，边走边追逐蟹类，会如拆机器零部件般将蟹类的足折断以便吞食，也吃螺类和沙虫、草地内的昆虫和蚯蚓。

# Eurasian Whimbrel

㊫ *Numenius phaeopus*

Protection categories：3V/LC

Status：Passing visitor

Body length：40–46cm

Habitat and Habits：Inhabit the intertidal zone of reefs，mudflats，beaches and coastal grasslands；like eating crabs and chasing them when walking；also feed on snails and sandworms，also insects and earthworms in the grassland.

bái yāo sháo yù

# 白腰杓鹬

# Eurasian Curlew

㊫ *Numenius arquata*

**保护等级：** 国二 / 近危

**居留类型：** 冬候鸟

**体长：** 57 ～ 63 厘米

**生活习性：** 常集大群活动各类海岸湿地生态环境，用其长喙插入泥沙中探测食物，偏好沙滩、泥沙滩等环境。主要取食沙滩、泥沙滩中下层的无脊椎动物等。

Protection categories: II/NT

Status: Winter visitor

Body length: 57–63cm

Habitat and Habits: Form large groups in coastal wetlands, and probe mud for food with long bill. Prefer sandy and muddy beaches. Feed on benthic invertebrates.

fān shí yù

# 翻石鹬

# Ruddy Turnstone

㊫ *Arenaria interpres*

**保护等级：** 国二 / 无危

**居留类型：** 过境鸟

**体长：** 21 ～ 26 厘米

**生活习性：** 结小群栖于沿海泥滩、沙滩及海岸石岩。有时在内陆或近海开阔处进食。在海滩上翻动石头及其他物体找食甲壳类。奔走迅速。

Protection categories: II/LC

Status: Passing visitor

Body length: 21–26cm

Habitat and Habits: Inhabit in groups on coastal mudflats, sandy beaches, and rocky shores. Sometimes forage inland or open areas along the coasts. Flip over stones and other objects for searching prey like crustacean. Move fast.

dà bīn yù
# 大滨鹬

# Great Knot

㊫ *Calidris tenuirostris*

**保护等级：**国二 / 濒危

**居留类型：**冬候鸟

**体长：**26 ～ 30 厘米

**生活习性：**常集大群或混群活动于海岸湿地生态环境，偏好沙滩和泥沙滩。喙触觉灵敏，食性较广，日夜取食滨海湿地各类底栖生物。

Protection categories: II / EN

Status: Winter visitor

Body length: 26–30cm

Habitat and Habits: Gather in groups at coastal wetlands, favoring sandy and muddy beaches. Feed on various benthic organisms day and night.

国家二级保护野生动物：大滨鹬

The national second-class protected wild animal: Great Knot

kuò zuǐ yù
# 阔嘴鹬

# Broad-billed Sandpiper

㊫ *Calidris falcinellus*

**保护等级：**国二 / 无危

**居留类型：**冬候鸟

**体长：**15 ～ 18 厘米

**生活习性：**喜潮湿的沿海泥滩、沙滩及沼泽地区。翻找食物时嘴垂直向下。遇到危险时蹲伏。

Protection categories：II/LC

Status：Winter visitor

Body length：15–18cm

Habitat and Habits：Prefers moist coastal mudflats, sandy beaches, and boggy areas. Forages by probing its bill downward into the mud. Takes a crouching position when alarmed.

jiān wěi bīn yù

# 尖尾滨鹬

# Sharp-tailed Sandpiper

㊫ *Calidris acuminata*

**保护等级：** 三有 / 无危

**居留类型：** 过境鸟

**体长：** 16～23 厘米

**生活习性：** 活动于海岸潮间带、河口、盐田、沼泽、农田、草地等生境。以螺贝、虾蟹、沙虫、昆虫及种子为食。

**Protection categories:** 3V/LC

**Status:** Passing visitor

**Body length:** 16–23cm

**Habitat and Habits:** Inhabit coastal intertidal zones, estuaries, salt pans, marshes, farmlands, grasslands, and other habitats; feed on snails, shrimps, crabs, sandworms, insects, and seeds.

wān zuǐ bīn yù

# 弯嘴滨鹬

# Curlew Sandpiper

㊫ *Calidris ferruginea*

**保护等级：** 三有 / 近危

**居留类型：** 冬候鸟

**体长：** 18 ～ 23 厘米

**生活习性：** 栖于沿海滩涂及近海的稻田和鱼塘。通常与其他滨鹬及鹬类混群。潮落时跑至泥里翻找食物。

**Protection categories:** 3V /NT

**Status:** Winter visitor

**Body length:** 18–23cm

**Habitat and Habits:** Inhabits coastal mudflats, rice fields, and fish ponds. Often seen mingling with other sandpipers and plovers. Forages by running and flipping mud for food during low tide.

qīng jiǎo bīn yù

# 青脚滨鹬

# Temminck's Stint

㊫ *Calidris temminckii*

**保护等级：** 三有 / 无危

**居留类型：** 冬候鸟

**体长：** 13～15 厘米

**生活习性：** 常见于淡水水域，也光顾潮间带泥滩，与其他小型涉禽混群。受惊时直接从地面快速起飞，成群做盘旋飞行。

Protection categories: 3V/LC

Status: Winter visitor

Body length: 13–15cm

Habitat and Habits: Often seen in freshwater areas, and mud in the intertidal area. Mingling with other small waders. When frightened, take off quickly from the ground and fly around in groups.

chánɡ zhǐ bīn yù

# 长趾滨鹬

# Long-toed Stint

㊫ *Calidris subminuta*

**保护等级：** 三有 / 无危

**居留类型：** 冬候鸟

**体长：** 13 ~ 16 厘米

**生活习性：** 偏好淡水环境，活动于池塘、水田、湖泊、沼泽等内陆淡水湿地。站姿较挺直，以昆虫、螺贝、种子为食。

**Protection categories:** 3V/LC

**Status:** Winter visitor

**Body length:** 13–16cm

**Habitat and Habits:** Prefer freshwater environments, and inhabit ponds, paddy fields, lakes, marshes, and other inland freshwater wetlands; stand upright; feed on insects, snails, and seeds.

sān zhǐ bīn yù

# 三趾滨鹬

保护等级：三有 / 无危

居留类型：冬候鸟

体长：19 ～ 21 厘米

生活习性：喜滨海沙滩，极少至泥地。通常随落潮在水边奔跑，同时捡食海潮冲刷出来的小食物。

# Sanderling

㊫ *Calidris alba*

Protection categories: 3V / LC

Status: Winter visitor

Body length: 19–21cm

Habitat and Habits: Prefers coastal sandy beaches, rarely ventures into muddy areas. Often seen running along the edge of waves picking up food items.

hóng jǐng bīn yù
# 红颈滨鹬

# Red-necked Stint

*Calidris ruficollis*

**保护等级：** 三有 / 近危

**居留类型：** 冬候鸟

**体长：** 13 ～ 16 厘米

**生活习性：** 喜集群活动于泥沙滩湿地或者河口地区，与其他小型鸻鹬类混群。觅食时在泥滩上边走边用喙从泥滩表层啄食，以沙虫、虾蟹、螺贝、昆虫及种子为食。

**Protection categories:** 3V / NT

**Status:** Winter visitor

**Body length:** 13–16cm

**Habitat and Habits:** Like to form flocks in mud flat or estuaries, and often mix with other small plovers; walk and peak at the surface of mudflats when foraging; feed on sandworms, shrimps, crabs, snails, insects, and seeds.

hēi fù bīn yù
# 黑腹滨鹬

保护等级：三有 / 无危

居留类型：冬候鸟

体长：16 ～ 22 厘米

生活习性：常集大群活动于沿海滩涂、河口、水田等多种湿地生境。觅食时分散，遇到危险会集成大群在空中不断翻飞转向，直到威胁解除。以甲壳类、螺贝、昆虫等无脊椎动物为食。

# Dunlin

㊫ *Calidris alpina*

Protection categories：3V / LC

Status：Winter visitor

Body length：16–22cm

Habitat and Habits：Often form large flocks in coastal beaches, estuaries, paddy fields and other wetland habitats; separate when foraging; integrate in large groups when in danger, and constantly fly around in the air until the threat is removed; feed on crustaceans, snails, insects, and other invertebrates.

qiū yù
# 丘鹬

# Eurasian Woodcock

㊫ *Scolopax rusticola*

**保护等级：** 三有 / 无危

**居留类型：** 冬候鸟

**体长：** 33 ~ 38 厘米

**生活习性：** 常单独活动于各类森林、灌木环境下阴暗的地面，主要在夜间活动，性隐秘。主要取食在地面上活动的各类无脊椎动物。

**Protection categories:** 3V/LC

**Status:** Winter visitor

**Body length:** 33–38cm

**Habitat and Habits:** Often solitary, they inhabit shadowed ground of forest and shrub, primarily nocturnal and secretive. Feed mainly on various invertebrates that inhabit the forest floor.

zhēn wěi shā zhuī
# 针尾沙锥

# Pin-tailed Snipe

㊫ *Gallinago stenura*

**保护等级：** 三有 / 无危

**居留类型：** 过境鸟

**体长：** 25 ～ 27 厘米

**生活习性：** 活动于稀疏水草的带点浅水的湿泥地，常单独活动，或略微分散的小群，惊飞时“Z”字形飞行。以蠕虫、昆虫、螺及种子为食。

**Protection categories:** 3V/LC

**Status:** Passing visitor

**Body length:** 25–27cm

**Habitat and Habits:** Inhabit shallow wet mudflats with sparse grass; often act alone or form small flocks; Fly in a Z-shaped pattern when startled; feed on worms, insects, snails, and seeds.

shàn wěi shā zhuī

# 扇尾沙锥

**保护等级：**三有 / 无危

**居留类型：**冬候鸟

**体长：**24 ～ 29 厘米

**生活习性：**活动于有水草的水田、沼泽、河岸、湖边等大面积湿地，惊飞时“Z”字形飞行。觅食时以喙垂直插入泥中，捕食蠕虫、昆虫、螺、小鱼。

# Common Snipe

学 *Gallinago gallinago*

**Protection categories:** 3V/LC

**Status:** Winter visitor

**Body length:** 24–29cm

**Habitat and Habits:** Inhabit paddy fields, mudflats, riversides, lakesides, and other large wetlands with a variety of vegetation; fly in a Z-shaped pattern when startled; vertically insert beaks into the mud when foraging; prey on worms, insects, snails, small fishes.

qiào zuǐ yù

# 翘嘴鹬

# Terek Sandpiper

㊫ *Xenus cinereus*

**保护等级：** 三有 / 无危

**居留类型：** 过境鸟

**体长：** 22 ～ 25 厘米

**生活习性：** 常集小群或混群活动于各类海岸生态环境，也可见于河口等湿地，偏好咸水、半咸水等地。视觉灵敏，跑动速度快，主要用喙啄食湿地表层的小型动物，也可用喙探测泥中食物。

Protection categories: 3V / LC

Status: Passing visitor

Body length: 22–25cm

Habitat and Habits: Often form flocks or mixed groups in variety of coastal areas, also seen in estuaries and other wetlands; prefer brackish water. Keen eyesight and fast runners. Use bills to peck surface prey and probe mud for food.

hóng jǐng bàn pǔ yù

# 红颈瓣蹼鹬

# Red-necked Phalarope

㊫ *Phalaropus lobatus*

**保护等级：** 三有 / 无危

**居留类型：** 过境鸟

**体长：** 16 ~ 20 厘米

**生活习性：** 冬季在海上结大群，食物为浮游生物。不甚惧人，易于接近。有时到陆上的池塘或沿海滩涂取食。

Protection categories: 3V/LC

Status: Passing visitor

Body length: 16-20cm

Habitat and Habits: In winter, form large flocks at sea, feeding on plankton. Fearless of humans and easily approachable. Sometimes forage in inland ponds or along coastal mudflats.

jī yù
# 矶鹬

# Common Sandpiper

㊫ *Actitis hypoleucos*

**保护等级：** 三有 / 无危

**居留类型：** 冬候鸟

**体长：** 16 ～ 22 厘米

**生活习性：** 活动于海岸、河口、沼泽及水田等湿地，通常单独活动，尾羽上下摆动，飞行时则贴着水面飞行。主食昆虫及其幼虫。

**Protection categories:** 3V / LC

**Status:** Winter visitor

**Body length:** 16–22cm

**Habitat and Habits:** Inhabit coasts, estuaries, mudflats, paddy fields and other wetlands; often inhabit alone; swing tail up and down and fly close to water; mainly prey on insects and larvae.

bái yāo cǎo yù
# 白腰草鹬

# Green Sandpiper

㊫ *Tringa ochropus*

**保护等级：** 三有 / 无危

**居留类型：** 冬候鸟

**体长：** 21 ～ 24 厘米

**生活习性：** 常单独活动于湖泊、河流等湿地生态环境下水草茂密的地方，常于泥草交接处。取食相应环境下各类无脊椎动物、小鱼和小虾等。

Protection categories: 3V/LC

Status: Winter visitor

Body length: 21–24cm

Habitat and Habits: Solitary in lake, river and other wetland habitat with dense aquatic vegetation, feeds on invertebrates and small fishes and shrimps.

huī wěi piāo yù

# 灰尾漂鹬

# Grey-tailed Tattler

㊫ *Tringa brevipes*

**保护等级：** 三有 / 近危

**居留类型：** 过境鸟

**体长：** 23 ～ 28 厘米

**生活习性：** 常单独或小群觅食于沿海滩涂或河口湿地，喜多石头的海岸。在礁石或泥滩上跑步觅食，偶尔在浅水中探索猎物，将食物在水边清洗后食入。以虾蟹、沙虫、螺贝、昆虫及小鱼为食。

**Protection categories:** 3V/NT

**Status:** Passing visitor

**Body length:** 23–28cm

**Habitat and Habits:** Often inhibit alone or form small groups in coastal beaches or estuary wetlands; prefer rocky shoreline, and forage on rocky or sandy substrates; sometimes prey in the shallow water areas, eat food after washing it; feed on shrimps, marine worms, snails, insects, and small fish.

lín yù
# 林鹬

# Wood Sandpiper

㊫ *Tringa glareola*

**保护等级：** 三有 / 无危

**居留类型：** 冬候鸟

**体长：** 19 ～ 23 厘米

**生活习性：** 偏好淡水环境，常集群于浅水草地中觅食，极少到达海边或湖口。食物以昆虫为主，也吃螺贝、甲壳类、蠕虫、小鱼及蛙类。

**Protection categories：** 3V / LC

**Status：** Winter visitor

**Body length：** 19–23cm

**Habitat and Habits:** Prefer freshwater environments, often form flocks and forage in shallow-water grassland, and rarely reach coasts or estuaries; mainly feed on insects, and also feed on snails, crustaceans, worms, small fish and frogs.

hóng jiǎo yù
# 红脚鹬

# Common Redshank

㊫ *Tringa totanus*

**保护等级：** 三有 / 无危

**居留类型：** 冬候鸟

**体长：** 26 ～ 29 厘米

**生活习性：** 越冬和迁徙期间常结小群活动于泥岸、海滩、盐田、干涸的沼泽、鱼塘及近海稻田，与其他涉禽混群。

**Protection categories:** 3V/LC

**Status:** Winter visitor

**Body length:** 26–29cm

**Habitat and Habits:** During winter and migration, forms small groups and forages in mudflats, beaches, salt pans, dried marshes, fish ponds, and nearshore rice fields, often mingling with other waders.

hè yù
# 鹤鹬

# Spotted Redshank

*Tringa erythropus*

**保护等级：**三有 / 无危

**居留类型：**冬候鸟

**体长：**26 ~ 33 厘米

**生活习性：**常集群活动于沿海滩涂、河口、池塘、水田、沼泽等湿地。在浅水中用喙左右扫动觅食，在较深水域则将头探入水中觅食，会游泳。以昆虫、虾蟹、螺贝、小鱼为食。

**Protection categories:** 3V/LC

**Status:** Winter visitor

**Body length:** 26–33cm

**Habitat and Habits:** Gather in groups along coastal beaches, estuaries, ponds, paddy fields, swamps and other wetlands. Sweeps bill in shallow water and dips head underwater in deeper areas, capable of swimming. Feeds on insects, crustaceans, mollusks, and small fish.

qīng jiǎo yù

# 青脚鹬

**保护等级：** 三有 / 无危
**居留类型：** 冬候鸟
**体长：** 30 ～ 35 厘米
**生活习性：** 常单独或小群活动于湿地泥滩，或与其他鸻鹬类混群觅食，惊飞或返回时集小群，飞行时常发出三音节“丢丢丢”的叫声。主要捕食小鱼或昆虫。

# Common Greenshank

㊫ *Tringa nebularia*

**Protection categories:** 3V/LC
**Status:** Winter visitor
**Body length:** 30–35cm
**Habitat and Habits:** Often inhabit alone or form small flocks in mudflat wetlands; forage in mixed groups with other sandpipers and plovers; often emit a three-note "tew-tew-tew" call during flight; mainly feed on small fish or insects.

# zé yù 泽鹬

# Marsh Sandpiper

㊫ *Tringa stagnatilis*

**保护等级：** 三有 / 无危

**居留类型：** 冬候鸟

**体长：** 22 ～ 26 厘米

**生活习性：** 常集大群或混群活动于各类湿地环境，偏好咸水和咸淡水。在浅水中可扫动觅食，在泥滩中可将喙插入泥中探测食物，也可以啄食。

**Protection categories:** 3V/LC

**Status:** Winter visitor

**Body length:** 22–26cm

**Habitat and Habits:** Gathers in large or mixed groups in various wetland habitats, prefers brackish and saline water. Feeds by wading in shallow water, probing mud, and also picking with its bill.

pǔ tōng yàn héng

# 普通燕鸻

保护等级：三有 / 无危

居留类型：冬候鸟

体长：24 ～ 28 厘米

生活习性：喜欢栖息于水田或者开阔草地，集群活动。常集小群于湿地上空来回飞翔。地面行走觅食或飞行如燕般空中张口捕食飞虫。

# Oriental Pratincole

㊫ *Glareola maldivarum*

Protection categories: 3V/LC

Status: Winter visitor

Body length: 24–28cm

Habitat and Habits: Inhabit paddy field or open grasslands in groups. Often seen flying near wetlands; catch insects like swallow or find foods on the ground.

hóng zuǐ ōu
# 红嘴鸥

# Black-headed Gull

*Chroicocephalus ridibundus*

**保护等级：** 三有 / 无危

**居留类型：** 冬候鸟

**体长：** 36 ～ 42 厘米

**生活习性：** 繁殖期喜欢植被茂盛的浅水湿地，冬季则喜欢河口附近的泥质或沙质。

Protection categories: 3V / LC

Status: Winter visitor

Body length: 36–42cm

Habitat and Habits: Prefers densely vegetated shallow wetlands during the breeding season, and muddy or sandy areas near estuary wetlands during winter.

hēi wěi ōu
# 黑尾鸥

# Black-tailed Gull

㊫ *Larus crassirostris*

**保护等级：** 三有 / 无危

**居留类型：** 冬候鸟

**体长：** 46 ～ 48 厘米

**生活习性：** 常集小群活动于河口或近海的湿地环境中。性凶猛，常驱赶其他鸟类，包括体形与其相当的其他鸥类。取食水体中鱼类、软体动物、水草等。

**Protection categories:** 3V/LC

**Status:** Winter visitor

**Body length:** 46–48cm

**Habitat and Habits:** Inhabits estuary or coastal wetlands. Aggressive, drives away other birds including other gulls at similar size. Feeds on fish, mollusks, and aquatic plants.

xiǎo hēi bèi yín ōu
# 小黑背银鸥

# Lesser Black-backed Gull

㊫ *Larus fuscus*

**保护等级：** 无危

**居留类型：** 冬候鸟

**体长：** 51 ~ 70 厘米

**生活习性：** 常集小群活动于海边湿地环境中，海边泥滩走动或水体中游动。性温驯，夏季也见于内陆湖泊。取食水体中鱼类、软体动物、水草等。

Protection categories: LC

Status: Winter visitor

Body length: 51–70cm

Habitat and Habits: Gathers in small groups in coastal wetlands, walks on muddy beaches or swims. Docile, also found in inland lakes. Feeds on fish, mollusks, and aquatic plants.

hóng zuǐ jù yàn ōu
# 红嘴巨燕鸥

**保护等级：**三有 / 无危
**居留类型：**冬候鸟
**体长：**48 ～ 55 厘米
**生活习性：**常集大群活动，主要活动于滨海湿地，如内湾泥滩、河口湿地等，夏季也可到达内陆湖泊等有泥滩的湿地。主要取食水生鱼类，常从空中砸入水中抓鱼。

# Caspian Tern

㊫ *Hydroprogne caspia*

Protection categories：3V / LC
Status：Winter visitor
Body length：48–55cm
Habitat and Habits：Gathers in large groups in coastal wetlands and estuaries, also found in inland bay with mudflats. Feeds on aquatic fish, diving from the air to catch them in the water.

bái é yàn ōu

# 白额燕鸥

# Little Tern

㊫ *Sternula albifrons*

**保护等级：** 三有 / 无危

**居留类型：** 夏候鸟

**体长：** 20 ~ 28 厘米

**生活习性：** 常集小群活动于各类淡水湿地或淡咸水湿地，如河口滩涂、大型内陆湖泊沼泽地等。飞行迅速，可短时间悬停空中，头朝下，俯冲入水觅食。

**Protection categories：** 3V/LC

**Status：** Summer visitor

**Body length：** 20–28cm

**Habitat and Habits：** Form small groups in freshwater or brackish wetlands, like estuarine mudflats and inland lakes. Flies fast, splash dive in water to catch food.

pǔ tōng yàn ōu

# 普通燕鸥

# Common Tern

学 *Sterna hirundo*

**保护等级：** 三有 / 无危

**居留类型：** 冬候鸟

**体长：** 30 ～ 38 厘米

**生活习性：** 常集小群活动于各种湿地环境，以淡水湿地为主，如河流、湖泊、沼泽、池塘等。飞行能力强，可低空飞行，在水中取食后迅速腾空复飞。

**Protection categories:** 3V/LC

**Status:** Winter visitor

**Body length:** 30–38cm

**Habitat and Habits:** Often found in small groups in variety of wetland habitats, favoring freshwater like rivers, lakes, marshes, and ponds. They have a buoyant, graceful flight and hover over water before plunging down for food.

huī chì fú ōu
# 灰翅浮鸥

# Whiskered Tern

㊫ *Chlidonias hybrida*

**保护等级：** 三有 / 无危

**居留类型：** 过境鸟

**体长：** 23 ～ 28 厘米

**生活习性：** 结群活动，常至离海岸数十千米的内陆，在漫水地和稻田上空觅食，取食时扎入浅水或低掠水面。

**Protection categories:** 3V/LC

**Status:** Passing visitor

**Body length:** 23–28cm

**Habitat and Habits:** Forming flocks, they forage in shallow terrestrial wetlands and irrigated croplands, with certain distant from the coast; diving into shallow water or flying low over water, skimming the surface for food.

bái chì fú ōu
# 白翅浮鸥

# White-winged Tern
㊫ *Chlidonias leucopterus*

**保护等级：** 三有 / 无危
**居留类型：** 过境鸟
**体长：** 20 ～ 25 厘米
**生活习性：** 栖息于内陆水生植物茂盛的湿地中，越冬在沿海、海港及河口，也至内陆稻田及沼泽觅食。

**Protection categories:** 3V / LC
**Status:** Passing visitor
**Body length:** 20–25cm
**Habitat and Habits:** Inhabit inland wetlands with abundant aquatic vegetation and in coastal areas, harbors, and estuary during winter. Also forage in inland rice fields and marshes.

ōu zuǐ zào ōu

# 鸥嘴噪鸥

# Gull-billed Tern

㊫ *Gelochelidon nilotica*

**保护等级：** 三有 / 无危

**居留类型：** 冬候鸟

**体长：** 35 ~ 38 厘米

**生活习性：** 常光顾沿海河口、潟湖及内陆淡水湖或咸水湖。常徘徊飞行，取食时通常轻掠水面或于泥地捕食甲壳类及其他猎物，很少潜入水中。

**Protection categories:** 3V/LC

**Status:** Winter visitor

**Body length:** 35–38cm

**Habitat and Habits:** Frequent found in coastal estuaries, lagoons, and inland freshwater or saltwater lakes. Hover in flight, feed on crustaceans and other prey, by skimming water or capturing prey on muddy ground, rarely dive into water.

shān bān jiū
# 山斑鸠

## Oriental Turtle Dove

㊫ *Streptopelia orientalis*

**保护等级：** 三有 / 无危

**居留类型：** 留鸟

**体长：** 28 ～ 36 厘米

**生活习性：** 喜有林生境。成对或结小群活动于开阔农耕区及村庄，取食于地面。

Protection categories: 3V / LC

Status: Resident bird

Body length: 28–36cm

Habitat and Habits: Prefers forest habitats. Engages in pair or small group; forages near open agricultural areas and villages, feeding on the ground.

huǒ bān jiū
# 火斑鸠

# Red Collared Dove

学 *Streptopelia tranquebarica*

**保护等级：**三有 / 无危

**居留类型：**过境鸟

**体长：**20～23 厘米

**生活习性：**活动于开阔的田野、果园、低山林缘地带。常成群栖息于电线或高大枯枝上，以种子或小昆虫为食。

**Protection categories:** 3V/LC

**Status:** Passing visitor

**Body length:** 20–23cm

**Habitat and Habits:** Inhabit open fields, orchards, and forest edge in the low altitudes; often form flocks on the power lines or tall dead branches; feed on seeds or insects.

zhū jǐng bān jiū
# 珠颈斑鸠

# Spotted Dove

㊫ *Spilopelia chinensis*

**保护等级：** 三有 / 无危

**居留类型：** 留鸟

**体长：** 27 ～ 30 厘米

**生活习性：** 常单独或成对活动于树高层，公园草地上常见，巢穴对环境要求不高，空调外机上甚至晾晒的衣物里都可以筑巢。于地面觅食种子或昆虫。

**Protection categories:** 3V / LC

**Status:** Resident bird

**Body length:** 27–30cm

**Habitat and Habits:** Often inhabit alone or in pairs in the high layers of trees or park grassland; the requirement for nesting environment is not high, and nests could be built outside air conditioning machines or even in hanging clothes; feed on insects or seeds on the ground.

lǜ chì jīn jiū
# 绿翅金鸠

# Emerald Dove

㊫ *Chalcophaps indica*

**保护等级：**三有 / 无危

**居留类型：**留鸟

**体长：**22 ～ 25 厘米

**生活习性：**常单独或成对活动于阔叶林底层或林缘，或于地面上行走觅食。以果实、种子、嫩芽或小虫为食。

**Protection categories:** 3V/LC

**Status:** Resident bird

**Body length:** 22–25cm

**Habitat and Habits:** Often engage in activities solitary or in pair under broadleaf forest or the forest edge, or walk on the ground to forage. Feed on fruits, seeds, tender shoots, or small insects.

国家二级保护野生动物：褐翅鸦鹃

The national second-class protected wild animal: Greater Coucal

hè chì yā juān
# 褐翅鸦鹃

# Greater Coucal

㊫ *Centropus sinensis*

**保护等级：** 国二 / 无危

**居留类型：** 留鸟

**体长：** 47～56 厘米

**生活习性：** 常单独活动于湿地边上的密林或灌木中。性胆怯，遇到干扰常迅速藏入灌丛。以昆虫为主食，偶尔也吃草籽和果实。

**Protection categories：** II / LC

**Status：** Resident bird

**Body length：** 47–56cm

**Habitat and Habits：** Often inhabit alone in dense forests or shrubs on the edge of wetlands; timid, and often quickly hide in shrubs when disturbed; mainly feed on insects, but occasionally feed on grass seeds and fruits.

hóng chì fèng tóu juān

# 红翅凤头鹃

# Chestnut-winged Cuckoo

㊫ *Clamator coromandus*

**保护等级：**三有 / 无危

**居留类型：**夏候鸟

**体长：**38 ~ 46 厘米

**生活习性：**活动于林相较好的低矮丘陵地区，常单独或成对活动。性隐秘，善飞行，常抓空中飞虫；巢寄生。主食昆虫，偶尔吃浆果。

**Protection categories:** 3V/LC

**Status:** Summer visitor

**Body length:** 38–46cm

**Habitat and Habits:** Inhabit alone or in pairs in low hilly areas with well vegetated forests; Often found alone or in pairs. Secluded; good at flying, and often catch flying insects; parasitic; mainly feed on insects, but sometimes feed on berries.

红翅凤头鹃

Chestnut-winged Cuckoo

xiǎo yā juān
# 小鸦鹃

# Lesser Coucal

㊫ *Centropus bengalensis*

**保护等级：** 国二 / 无危

**居留类型：** 留鸟

**体长：** 34 ～ 38 厘米

**生活习性：** 栖息于山边或近水的灌丛和高草地中，常在地面活动，有时紧贴植被上方做短距离的飞行，并时常可见立于突出处，两翅半张开保持平衡。

Protection categories: II/LC

Status: Resident bird

Body length: 34–38cm

Habitat and Habits: Found in shrubs and tall grassy areas near hillsides or water. Active on the ground, sometimes fly close above vegetation for short distances. Often perches in a prominent object, and partially open wings for balance.

zào juān
# 噪鹃

# Asian Koel

学 *Eudynamys scolopaceus*

**保护等级：** 三有 / 无危

**居留类型：** 夏候鸟

**体长：** 39 ～ 46 厘米

**生活习性：** 活动于低海拔的次生林、公园、红树林等；巢寄生。以果实、种子为食，也捕食昆虫。

**Protection categories:** 3V/LC

**Status:** Summer visitor

**Body length:** 39–46cm

**Habitat and Habits:** Inhabit low-altitude secondary forests, parks, mangroves and other habitats; parasitic; feed on fruits, seeds and insects.

bā shēng dù juān

# 八声杜鹃

# Plaintive Cuckoo

㊫ *Cacomantis merulinus*

**保护等级：** 三有 / 无危

**居留类型：** 夏候鸟

**体长：** 21 ～ 25 厘米

**生活习性：** 喜开阔林地、次生林及农耕区，包括城镇和村庄周围树丛，隐匿于树冠层。利用长尾缝叶莺代为孵卵。

**Protection categories:** 3V/LC

**Status:** Summer visitor

**Body length:** 21–25cm

**Habitat and Habits:** Prefer open woodland, secondary forest, and farmland, including surrounding forest and villages; hidden in the canopy. Use Common Tailor bird to incubate eggs.

dà yīng juān

# 大鹰鹃

# Large Hawk-Cuckoo

㊫ *Hierococcyx sparverioides*

**保护等级：** 三有 / 无危

**居留类型：** 夏候鸟

**体长：** 38 ~ 42 厘米

**生活习性：** 常单独或成对活动于茂密的林层高层，善鸣，行为隐秘，常闻其声不见其影，巢寄生。主要以昆虫为食。

**Protection categories：** 3V/LC

**Status：** Summer visitor

**Body length：** 38–42cm

**Habitat and Habits：** Solitary or in pairs, inhabits dense upper canopy of forests. Elusive, known for vocalizations, often heard but rarely seen. Brood parasites and primarily feeds on insects.

sì shēng dù juān

# 四声杜鹃

# Indian Cuckoo

㊫ *Cuculus micropterus*

**保护等级：** 三有 / 无危

**居留类型：** 夏候鸟

**体长：** 31 ～ 34 厘米

**生活习性：** 常单独或成对活动。性隐秘，常隐藏于密林高处鸣唱；巢寄生。喜吃昆虫和浆果。

**Protection categories：** 3V/LC

**Status：** Summer visitor

**Body length：** 31–34cm

**Habitat and Habits：** Often inhabit alone or in pairs; secluded; often hide in the high layers of trees and sing; parasitic; prefer feeding on insects and berries.

zhōng dù juān

# 中杜鹃

# Himalayan Cuckoo

㊫ *Cuculus saturatus*

**保护等级：** 三有 / 无危

**居留类型：** 过境鸟

**体长：** 25 ~ 34 厘米

**生活习性：** 常单独活动于茂密的林层高层，善鸣，行为隐秘，迁徙季节遇见率较高，巢寄生。主要以昆虫为食。

**Protection categories:** 3V/LC

**Status:** Passing visitor

**Body length:** 25–34cm

**Habitat and Habits:** Solitary in dense forest canopy; known for their vocal; elusive behavior; commonly seen during migration, brood parasites, primarily feeds on insects.

国家二级保护野生动物：领角鸮
The national second-class protected wild animal: Collared Scops Owl

lǐng jiǎo xiāo

# 领角鸮

# Collared Scops Owl

㊫ *Otus lettia*

**保护等级：** 国二 / 无危

**居留类型：** 留鸟

**体长：** 23 ~ 25 厘米

**生活习性：** 常单独或成对活动，夜行性，常见于竹林或印度橡胶榕茂密的林子中间，晚上常鸣叫。主要以鼠类为食，也取食大型昆虫。

Protection categories: II/LC

Status: Resident bird

Body length: 23–25cm

Habitat and Habits: Solitary or in pairs, nocturnal, found in bamboo groves or dense forests, vocal at night. Feeds on rats, mice and large insects.

cǎo xiāo
# 草鸮

# Eastern Grass Owl

㊫ *Tyto longimembris*

**保护等级：**国二 / 无危

**居留类型：**留鸟

**体长：**32 ~ 38 厘米

**生活习性：**常单独活动于开阔的芒草地，夜行性，飞行时利用空气能力强。筑巢于高草堆中，巢穴两个出口，一个方便进出，另外一个方便逃生。主要以鼠类为食。

Protection categories: II/LC

Status: Resident bird

Body length: 32–38cm

Habitat and Habits: Solitary in open grasslands; nocturnal, strong flying ability, nests in grassy areas with two exits, one for entering and exiting, the other is for escaping; primarily feeds on rats and mice.

lín yè yīng
# 林夜鹰

# Savanna Nightjar
*Caprimulgus affinis*

**保护等级：** 三有 / 无危
**居留类型：** 留鸟
**体长：** 20 ～ 26 厘米
**生活习性：** 夜行性，多见于林地或郊野泥地路面，黄昏时活跃，主要以昆虫为食。白天多栖息于地上。

**Protection categories：** 3V / LC
**Status：** Resident bird
**Body length：** 20–26cm
**Habitat and Habits：** Nocturnal; mostly seen in woodland or mud pavement; active at dusk; mainly feed on insects; resting on ground during daytime.

xiǎo bái yāo yǔ yàn

# 小白腰雨燕

**保护等级：** 三有 / 无危

**居留类型：** 留鸟

**体长：** 11 ～ 15 厘米

**生活习性：** 对环境不挑剔，或高海拔或低海拔，或城市或荒野，常集小群或大群在空中来回飞行，也会与其他雨燕混群。飞行时张口捕食空中飞蚊、白蚁、蛾等小昆虫。

# House Swift

㊫ *Apus nipalensis*

**Protection categories:** 3V / LC

**Status:** Resident bird

**Body length:** 11–15cm

**Habitat and Habits:** Not picky to the environments; form small or large flocks in low or high altitudes, urban or wilderness; fly around in groups, sometimes mix with other swifts; prey on mosquitos, termites, moths, and other insects in the air.

sān bǎo niǎo

# 三宝鸟

**保护等级：** 三有 / 无危

**居留类型：** 夏候鸟

**体长：** 26 ～ 32 厘米

**生活习性：** 常栖于近林开阔地的枯树上，偶尔起飞追捕过往昆虫，或向下俯冲捕捉地面昆虫。

# Oriental Dollarbird

㊫ *Eurystomus orientalis*

**Protection categories:** 3V / LC

**Status:** Summer visitor

**Body length:** 26–32cm

**Habitat and Habits:** Perches on dead trees in open areas near forests, occasionally takes flight to chase insects or dives down to capture ground-dwelling insects.

lì hóu fēng hǔ
# 栗喉蜂虎

# Blue-tailed Bee-eater

*Merops philippinus*

**保护等级：** 国二 / 无危
**居留类型：** 过境鸟
**体长：** 25 ~ 36 厘米
**生活习性：** 集群活动于海边开阔的林边，偏好在沙土或裸露的小山坡挖洞为巢，常停栖于电线或裸露的树枝。飞行捕食空中的蜜蜂、蝴蝶、飞蛾等昆虫。

Protection categories: II/LC
Status: Passing visitor
Body length: 25–36cm
Habitat and Habits: Often seen in groups inhabiting open forage edge; prefer to dig holes in sandy soil or exposed hillsides for nests; often rest on the power lines or bare branches; floppily glide in the air, and prey on bees, butterflies, moths, and other insects.

lán fěi cuì
# 蓝翡翠

# Black-capped Kingfisher

*Halcyon pileata*

**保护等级：** 三有 / 无危

**居留类型：** 留鸟

**体长：** 26 ～ 31 厘米

**生活习性：** 喜大河流两岸，非繁殖季节也见于河口及红树林。栖于悬于河上的枝头。

Protection categories: 3V/LC

Status: Resident bird

Body length: 26–31 cm

Habitat and Habits: Prefers riverbanks; can be found in estuaries and mangroves during non-breeding season. Perches on branches hanging over the river.

bái xiōng fěi cuì

# 白胸翡翠

# White-throated Kingfisher

㊫ *Halcyon smyrnensis*

**保护等级：** 国二 / 无危

**居留类型：** 留鸟

**体长：** 26 ～ 29 厘米

**生活习性：** 常单独活动于开阔的水库、沼泽、稻田等地，繁殖期可见于城市，凌晨常在建筑物高处鸣叫。捕食鱼类、两栖类、昆虫和小鼠。

Protection categories: II/LC

Status: Resident bird

Body length: 26–29cm

Habitat and Habits: Often inhabit alone in open reservoirs, swamps, rice paddies and other habitats; could be seen in cities when breeding, and often call in early morning at the height of buildings; prey on fish, amphibians, insects, rats and mice.

pǔ tōng cuì niǎo

# 普通翠鸟

# Common Kingfisher

*Alcedo atthis*

**保护等级：** 三有 / 无危

**居留类型：** 留鸟

**体长：** 15 ~ 17 厘米

**生活习性：** 常单独活动于各种浅水区域，海边湿地、湖泊、河流、农田等。常伫立于水边枝头上等待猎物的出现，迅速冲入水中捕食，或空中定点振翅悬停，再俯冲入水捉鱼。

**Protection categories:** 3V / LC

**Status:** Resident bird

**Body length:** 15–17cm

**Habitat and Habits:** Often inhabit shallow water, coastal wetland, lake, river, paddy fields, etc. Stand on the branches close to water and wait for preys, then quickly rushing into the water to prey; or hovering in the air, and then diving into the water to catch fish.

bān yú gǒu

# 斑鱼狗

## Pied Kingfisher

㊫ *Ceryle rudis*

**保护等级：** 三有 / 无危

**居留类型：** 留鸟

**体长：** 27 ～ 31 厘米

**生活习性：** 成对或结群活动于较大水体及红树林。唯一常盘桓水面寻食的鱼狗。

**Protection categories:** 3V/LC

**Status:** Resident bird

**Body length:** 27–31cm

**Habitat and Habits:** Pairs or groups found near water bodies and mangroves. Usually hunt by hovering over the water to detect prey.

dài shèng

# 戴胜

# Common Hoopoe

*Upupa epops*

**保护等级：**三有 / 无危

**居留类型：**留鸟

**体长：**25 ~ 31 厘米

**生活习性：**常单独或小群活动于草地、海边开阔林地。飞行时翅膀一张一合，动作优美。头上冠羽张开有警戒、求偶或平衡功能。快速点头式挖掘土壤中的虫或虫卵蛹为食。

**Protection categories：**3V/LC

**Status：**Resident bird

**Body length：**25–31cm

**Habitat and Habits：**Often inhabit alone or in small group in grasslands, and open forests in the coasts; close and open wings when flying, with beautiful movement; outspread head feathers with warming, mating, and balancing functions; probe the ground for worms, insect eggs or pupa with a quick nod.

yǐ liè
# 蚁䴕

# Eurasian Wryneck

*Jynx torquilla*

**保护等级：** 三有 / 无危

**居留类型：** 冬候鸟

**体长：** 16 ～ 19 厘米

**生活习性：** 单独活动为主，常活动于开阔稀疏的林间，飞行呈波浪状。主要在地面觅食蚂蚁。

Protection categories: 3V/LC

Status: Winter visitor

Body length: 16–19cm

Habitat and Habits: Mainly inhabit alone in open forests; flight is somewhat undulating; mainly prey on the ground for ants.

hóng sǔn

# 红隼

# Common Kestrel

㊫ *Falco tinnunculus*

**保护等级：** 国二 / 无危

**居留类型：** 冬候鸟

**体长：** 31 ～ 38 厘米

**生活习性：** 常活动于多草和低矮植被的开阔地带，停栖于电线、树桩、枯枝等显眼位置。飞行时经常定点振翅悬停觅食，喜食大型昆虫与鼠类。

**Protection categories：** II/LC

**Status：** Winter visitor

**Body length：** 31–38cm

**Habitat and Habits：** Often found in open areas with grass and shrubs; perches on visible spots like power lines, tree stumps and bare branches. Hovers in flight to prey, feeding on large insects, rats and mice.

hóng jiǎo sǔn
# 红脚隼

# Amur Falcon

㊫ *Falco amurensis*

**保护等级：** 国二 / 无危

**居留类型：** 过境鸟

**体长：** 25 ～ 30 厘米

**生活习性：** 迁徙季节集大群，以迁徙为主，或单一种群，或与其他迁徙猛禽混群，傍晚时也停歇于电线上，利用空气能力强。喙较弱，主要以飞行的昆虫等为食。

Protection categories: II/LC

Status: Passing visitor

Body length: 25–30cm

Habitat and Habits: During migration, they form large flocks, with their own species, or with other raptors; roost on power lines, exhibit powerful aerial abilities in flight; and feed on flying insects.

yàn sǔn

# 燕隼

# Eurasian Hobby

㊫ *Falco subbuteo*

**保护等级：** 国二 / 无危

**居留类型：** 过境鸟

**体长：** 29 ～ 35 厘米

**生活习性：** 迁徙季节集大群，常与其他过境猛禽混群迁徙，利用空气能力超强。主要以飞行的昆虫等为食，也能捕捉空中飞行的燕子等。

**Protection categories：** II/LC

**Status：** Passing visitor

**Body length：** 29–35cm

**Habitat and Habits：** Forms large flocks and exhibit powerful aerial abilities in flight during migration; mainly feeds on flying insects and can also capture birds in flight.

yóu sǔn
# 游隼

# Peregrine Falcon

㊫ *Falco peregrinus*

**保护等级：**国二 / 无危

**居留类型：**留鸟

**体长：**41 ～ 50 厘米

**生活习性：**常单独活动，有迁徙个体，以留鸟为主，体形相对较大，利用空气能力强，收翼俯冲时速度极快，可撞晕比其体形大得多的鸟类，如野鸭等。肉食性，主要以鸟类为食，也捕食其他两栖类。

Protection categories：II/LC

Status：Resident bird

Body length：41–50cm

Habitat and Habits：Solitary, mainly residents, with migratory individual. Large body size with strong aerial abilities in flight. Dives at high speeds, stunning larger birds, like ducks. Primarily feeds on birds and also prey on amphibians.

xiān bā sè dōng

# 仙八色鸫

# Fairy Pitta

㊫ *Pitta nympha*

**保护等级：** 国二 / 易危

**居留类型：** 过境鸟

**体长：** 16 ～ 20 厘米

**生活习性：** 常单独或成对活动于林相较好的林地底层及灌丛中，尤其是林中小溪边或者林中水田边。性隐秘，可长时间站立不动，迁徙时城市公园可见。主要以落叶下方松土中无脊椎动物等为食，如蚯蚓等。

**Protection categories:** II / VU

**Status:** Passing visitor

**Body length:** 16–20cm

**Habitat and Habits:** Often solitary or in pairs, found in the dense ground in shrubs of well-forested areas, especially near streams or paddy field. Secretive, and stand still for a long time. Primarily feeds on soil-dwelling invertebrates such as earthworms beneath fallen leaves.

huī hóu shān jiāo niǎo

# 灰喉山椒鸟

# Grey-chinned Minivet

㊫ *Pericrocotus solaris*

**保护等级：** 三有 / 无危

**居留类型：** 留鸟

**体长：** 17 ～ 19 厘米

**生活习性：** 常集小群活动于山坡树林树冠层。性活跃，飞行时常常伴随鸣叫声，常于林缘或沿着路边来回活动。飞行能力强，采取集体活动策略，飞行惊动树中昆虫，为同伴捕食提供便利。

**Protection categories:** 3V / LC

**Status:** Resident bird

**Body length:** 17–19cm

**Habitat and Habits:** Gathers in small groups in forest canopy, exhibiting active behavior and making a twittering call while flying. Strong flying abilities. Adopt a collective foraging strategy to startle insects, facilitating hunting by companions.

雄鸟 ♂

雌鸟 ♀

chì hóng shān jiāo niǎo

# 赤红山椒鸟

# Scarlet Minivet

㊫ *Pericrocotus speciosus*

**保护等级：**三有 / 无危

**居留类型：**留鸟

**体长：**17 ～ 22 厘米

**生活习性：**行为似灰喉山椒鸟，常集群或混群活动于山林树冠层。性活跃，飞行时常常伴随鸣叫声，常于林缘或沿着路边来回活动。集体觅食，互相通过飞行惊动昆虫，为同伴觅食提供便利。

**Protection categories:** 3V/LC

**Status:** Resident bird

**Body length:** 17–22cm

**Habitat and Habits:** Similar to Grey-chinned Minivet. Active in flocks or mixed-speices groups in forest canopy, making a twittering call while flying, often seen along forest edges. Engage in collective foraging, startling insects to aid companions in finding food.

àn huī juān jú

# 暗灰鹃䴗

保护等级：三有 / 无危

居留类型：过境鸟

体长：19 ～ 24 厘米

生活习性：单独或成对活动于林间，有时也与其他鸟混群，多活动于树冠层。捕食昆虫。

# Black-winged Cuckooshrike

㊫ *Lalage melaschistos*

Protection categories: 3V/LC

Status: Passing visitor

Body length: 19-24cm

Habitat and Habits: Often inhabit alone or in pairs in forests, or sometimes mix with other birds; mostly act in the canopy, and prey on insects.

hóng wěi bó láo

# 红尾伯劳

**保护等级：**三有 / 无危

**居留类型：**冬候鸟

**体长：**17 ～ 20 厘米

**生活习性：**常单独活动于开阔的农田及灌木，领地性较强，常驱赶同类。捕食昆虫、蛙类、小型鸟类、幼鼠等。

# Brown Shrike

学 *Lanius cristatus*

**Protection categories:** 3V/LC

**Status:** Winter visitor

**Body length:** 17–20cm

**Habitat and Habits:** Often inhabit alone in open fields and shrubs; strongly territorial and often drive away other species; prey on insects, frogs, small birds, rats and mice, etc.

zōng bèi bó láo

# 棕背伯劳

# Long-tailed Shrike

学 *Lanius schach*

**保护等级：** 三有 / 无危

**居留类型：** 留鸟

**体长：** 20 ~ 25 厘米

**生活习性：** 性凶猛，肉食性，领域性强，常单独站立于高处鸣叫以宣示领地。捕食昆虫、蛙类、小型鸟类，常把猎物挂于树枝储藏食物。

**Protection categories:** 3V/LC

**Status:** Resident bird

**Body length:** 20–25cm

**Habitat and Habits:** Fierce, carnivorous, and strongly territorial; often stand alone on a height to call for territory declaration; prey on insects, frogs, and small birds, and often hang their preys on branches to store food.

hēi zhěn huáng lí

# 黑枕黄鹂

**保护等级**：三有 / 无危

**居留类型**：过境鸟

**体长**：23～28 厘米

**生活习性**：栖于开阔林、人工林、园林、村庄及红树林。成对或以家族为群活动。常留在树上但有时下至低处捕食昆虫。飞行呈波状，振翼幅度大，缓慢而有力。

# Black-naped Oriole

㊫ *Oriolus chinensis*

**Protection categories**: 3V / LC

**Status**: Passing visitor

**Body length**: 23–28cm

**Habitat and Habits**: Inhabit open forests, plantations, gardens, villages, and mangroves. Seen in pairs or family groups. Mainly stay in trees but occasionally descend to prey on insects.

成鸟 ad.

huī juǎn wěi
# 灰卷尾

**保护等级：**三有 / 无危
**居留类型：**过境鸟
**体长：**26 ～ 28 厘米
**生活习性：**活动于林缘或林间，常站立于突出的枝头，迅速在空中捕食昆虫，领地性较强，会攻击猛禽等入侵者。

# Ashy Drongo

㊫ *Dicrurus leucophaeus*

Protection categories：3V/LC
Status：Passing visitor
Body length：26–28cm
Habitat and Habits：Inhabit forage edge or inside forest, often stand on the protruding branches; quickly hunt insects in air; strongly territorial, and will attack raptors and other intruders.

hēi juǎn wěi

# 黑卷尾

# Black Drongo

学 *Dicrurus macrocercus*

**保护等级：** 三有 / 无危

**居留类型：** 夏候鸟

**体长：** 24 ～ 30 厘米

**生活习性：** 常集群活动于开阔而稀疏的林地，如农田边的树上、大草坪边的枝头等。主食昆虫，繁殖期性情凶猛，偶见捕食其他鸟类，有时候也会攻击猛禽。

**Protection categories:** 3V/LC

**Status:** Summer visitor

**Body length:** 24–30cm

**Habitat and Habits:** Often form flocks in open forests, such as on the trees near the fields, or branches on the edge of lawns; mainly prey on insects; fierce when breeding; sometimes prey on other birds, and attack raptors.

fà guān juǎn wěi

# 发冠卷尾

**保护等级：** 三有 / 无危

**居留类型：** 夏候鸟

**体长：** 29 ～ 34 厘米

**生活习性：** 常于茂密的林间活动，转弯能力强，能在密林中穿梭，捕食林间飞虫。

# Hair-crested Drongo

㊫ *Dicrurus hottentottus*

**Protection categories:** 3V/LC

**Status:** Summer visitor

**Body length:** 29-34cm

**Habitat and Habits:** Inhabit alone or in pairs in forests and can fly through in dense forest; prey on insects in air.

hēi zhěn wáng wēng

# 黑枕王鹟

# Black-naped Monarch

㊫ *Hypothymis azurea*

**保护等级：**无危

**居留类型：**留鸟

**体长：**14 ~ 16 厘米

**生活习性：**常单独或成对活动于树林中，主要活动于树冠层，极少下地，迁徙季节可见于城市公园的树上。呈波浪形飞行，动作优美。取食树中昆虫成虫及大型幼虫。

**Protection categories:** LC

**Status:** Resident bird

**Body length:** 14-16cm

**Habitat and Habits:** Often solitary or in pairs, mainly inhabits the forest canopy, rarely descending to the ground. During migration, they can be seen in urban parks. They feed on insects and larvae in trees.

shòu dài
# 寿带

# Amur Paradise Flycatcher

㊫ *Terpsiphone incei*

**保护等级：** 三有 / 无危

**居留类型：** 过境鸟

**体长：** 20 ～ 42 厘米

**生活习性：** 常单独或成对活动于林间中下层，善飞行。以捕食飞行中的昆虫为主。

**Protection categories:** 3V/LC

**Status:** Passing visitor

**Body length:** 20–42cm

**Habitat and Habits:** Often inhabit alone or in pairs in the middle and lower layers of forests; strong flying abilities; mainly catching flying insects for food.

hóng zuǐ lán què
# 红嘴蓝鹊

# Red-billed Blue Magpie

㊫ *Urocissa erythroryncha*

**保护等级：** 三有 / 无危

**居留类型：** 留鸟

**体长：** 42 ～ 60 厘米

**生活习性：** 性喧闹，结小群活动。以果实、小型鸟类及卵、昆虫和动物尸体为食，常在地面取食。主动围攻猛禽。

**Protection categories:** 3V/LC

**Status:** Resident bird

**Body length:** 42–60cm

**Habitat and Habits:** Noisy and social, forms small groups. Feeds on fruits, small birds, eggs, insects, and carrion, and often forage on the ground. May actively attacks raptors.

xǐ què
# 喜鹊

# Oriental Magpie

㊫ *Pica serica*

**保护等级：**三有 / 无危

**居留类型：**留鸟

**体长：**40 ~ 50 厘米

**生活习性：**常单独或小群活动，各种生境都有，更偏好附近有大面积湿地生境的林间。杂食性，以昆虫、鼠类、小型鸟类及鸟蛋为食，也吃果实及种子。

**Protection categories:** 3V/LC

**Status:** Resident bird

**Body length:** 40–50cm

**Habitat and Habits:** Often inhabit alone or in small group in different habitats, and mostly prefer forest near wetland; omnivorous; prey on insects, rats, mice, small birds, bird eggs, and also feed on fruits and seeds

dà zuǐ wū yā
# 大嘴乌鸦

# Large-billed Crow

㊫ *Corvus macrorhynchos*

**保护等级：**无危

**居留类型：**留鸟

**体长：**45 ～ 47 厘米

**生活习性：**多成对或集小群栖于山地、丘陵、平原的树林中，非繁殖季节见于农田和村落。

Protection categories：LC

Status：Resident bird

Body length：45–47cm

Habitat and Habits：Often found in pair or in small groups in forests of mountains, hills and plains; in farmland and village during non-breading season.

dà shān què
# 大山雀

# Japanese Tit

㊫ *Parus minor*

**保护等级：** 三有

**居留类型：** 留鸟

**体长：** 12 ～ 14 厘米

**生活习性：** 单独或小群活动于林间中层，善握树枝，能爬树。以小型昆虫等无脊椎动物和花、嫩叶为食。

**Protection categories:** 3V

**Status:** Resident bird

**Body length:** 12–14cm

**Habitat and Habits:** Inhabit alone or in small groups in the middle layer of forest; good at holding branches, and hunting in trees; feed on small insects and other invertebrates, flowers and tender leaves.

zhōng huá pān què
# 中华攀雀

# Chinese Penduline Tit

㊫ *Remiz consobrinus*

**保护等级：** 三有 / 无危

**居留类型：** 冬候鸟

**体长：** 10 ～ 11 厘米

**生活习性：** 常活动于湿地芦苇、香蒲或高草地。以昆虫为主食，也吃草籽、浆果和嫩芽。

Protection categories: 3V/LC

Status: Winter visitor

Body length: 10–11cm

Habitat and Habits: Often inhabit wetland reed, cattails, or tall grassland; mainly prey on insects, and also feed on seeds, berries and tender leaves.

lì bèi duǎn jiǎo bēi

# 栗背短脚鹎

# Chestnut Bulbul

㊫ *Hemixos castanonotus*

**保护等级：**三有 / 无危

**居留类型：**留鸟

**体长：**19 ～ 21 厘米

**生活习性：**栖息于低山丘陵地区的次生阔叶林、林缘灌丛和稀树灌丛中。

**Protection categories:** 3V/LC

**Status:** Resident bird

**Body length:** 19–21cm

**Habitat and Habits:** Inhabits secondary broadleaf forests, shrubs at the edge of forests, and sparse woodland areas in lowland and foothills.

hēi duǎn jiǎo bēi
# 黑短脚鹎

# Black Bulbul

㊫ *Hypsipetes leucocephalus*

**保护等级：** 三有 / 无危

**居留类型：** 留鸟

**体长：** 23 ～ 26 厘米

**生活习性：** 常集小群或大群活动于山林中。性高调，常站立于树冠层的枝条上。采取集体觅食策略，互相通过飞行惊动昆虫，为同伴捕食昆虫提供帮助。

**Protection categories：** 3V / LC

**Status：** Resident bird

**Body length：** 23–26cm

**Habitat and Habits：** Gather in groups in mountain forests, exhibiting conspicuous behavior and perching on branches. Employs collective foraging strategies, using flight to flush out insects and assist companions.

bái tóu bēi
# 白头鹎

# Light-vented Bulbul

㊫ *Pycnonotus sinensis*

**保护等级：** 三有 / 无危

**居留类型：** 留鸟

**体长：** 18 ~ 20 厘米

**生活习性：** 活跃于各种生境，常小群活动，早晚间常活动于树冠层。性嘈杂。以果实、昆虫为食。

**Protection categories:** 3V/LC

**Status:** Resident bird

**Body length:** 18–20cm

**Habitat and Habits:** Inhabit various habitats, often form small groups, and act in crown layers in the morning and evening; noisy; prey on fruit and insects.

hóng ěr bēi

# 红耳鹎

# Red-whiskered Bulbul

㊫ *Pycnonotus jocosus*

**保护等级：** 三有 / 无危

**居留类型：** 留鸟

**体长：** 18 ～ 21 厘米

**生活习性：** 常活动于人类活动密集的公园环境。性活跃，筑巢于低矮的隐秘灌木中，常到地上活动。以果实、昆虫为食。

Protection categories: 3V/LC

Status: Resident bird

Body length: 18–21cm

Habitat and Habits: Often inhabit urban parks; active; build nests in low, hidden shrubs, and often act in the ground; prey on fruits and insects.

bái hóu hóng tún bēi

# 白喉红臀鹎

# Sooty-headed Bulbul

㊫ *Pycnonotus aurigaster*

**保护等级：**三有 / 无危

**居留类型：**留鸟

**体长：**19～21 厘米

**生活习性：**常见于开阔的城郊或者荒野地区，集小群活动，通常在树上觅食，偶尔到地上活动。以昆虫、果实、种子为食。

Protection categories: 3V/LC

Status: Resident bird

Body length: 19–21cm

Habitat and Habits: Often form small flocks in open suburban or wildness areas; forage in the trees, and sometimes act on the ground; prey on the insects, fruits and seeds.

jiā yàn
# 家燕

# Barn Swallow

㊫ *Hirundo rustica*

**保护等级：** 三有 / 无危

**居留类型：** 夏候鸟

**体长：** 17 ～ 19 厘米

**生活习性：** 亲近人类的鸟类，常集群飞行于农田上空，筑巢于屋檐下，巢穴开口向上。以蚊子等飞虫为食。

**Protection categories:** 3V / LC

**Status:** Summer visitor

**Body length:** 17–19cm

**Habitat and Habits:** Close to human habitation; forage in flocks flying above farmland; make nests under eaves, nests open upwards; feed on mosquitoes and other flying insects.

jīn yāo yàn
# 金腰燕

保护等级：三有 / 无危
居留类型：冬候鸟
体长：16 ～ 20 厘米
生活习性：主要栖息于低山丘陵和平原地区的村庄、城镇等居民区。

# Red-rumped Swallow

学 *Cecropis daurica*

Protection categories：3V/LC
Status：Winter visitor
Body length：16–20cm
Habitat and Habits：Primarily inhabits villages, towns, and residential areas in lowland and plain regions.

lì tóu zhī yè yīng

# 栗头织叶莺

**保护等级：** 三有 / 无危

**居留类型：** 冬候鸟

**体长：** 12 ～ 13 厘米

**生活习性：** 常单独或成对活动于山林中，声音尖锐但行为隐秘，领域性强，秋冬季节到海拔较低的环境觅食。主要捕食各类昆虫的幼虫。

# Mountain Tailorbird

㊫ *Phyllergates cucullatus*

ProtProtection categories：3V/LC

Status：Winter visitor

Body length：12–13cm

Habitat and Habits：Solitary or in pairs, found in mountain forests. Has a sharp call and exhibit secretive behavior. Forages in lower-altitude environments, primarily feeding on insect larvae.

yuǎn dōng shù yīng

# 远东树莺

# Manchurian Bush Warbler

㊫ *Horornis canturians*

**保护等级：** 无危

**居留类型：** 冬候鸟

**体长：** 15 ～ 18 厘米

**生活习性：** 常单独活动于旱地灌丛，秋冬季也可见于水边或农田边上的灌丛或者高草丛中。主要以小型节肢动物为食，如昆虫的幼虫、蜘蛛等。

**Protection categories:** LC

**Status:** Winter visitor

**Body length:** 15–18cm

**Habitat and Habits:** Solitary, found in dryland shrubs. In autumn and winter, can be seen in shrubs or tall grassy areas near water or farmland. Primarily feeds on small arthropods, such as insect larvae and spiders.

huáng méi liǔ yīng

# 黄眉柳莺

# Yellow-browed Warbler

㊫ *Phylloscopus inornatus*

**保护等级：** 三有 / 无危

**居留类型：** 冬候鸟

**体长：** 10 ～ 11 厘米

**生活习性：** 繁殖于高山森林中，迁徙时可见于低山次生林、城市绿化带、公园、果园及红树林等地。

**Protection categories:** 3V / LC

**Status:** Winter visitor

**Body length:** 10–11 cm

**Habitat and Habits:** Breeding in high mountain forests, can be observed during migration in low mountain secondary forests, urban green areas, parks, orchards, and mangroves.

huáng yāo liǔ yīng

# 黄腰柳莺

# Pallas's Leaf Warbler

㊫ *Phylloscopus proregulus*

**保护等级：** 三有 / 无危

**居留类型：** 冬候鸟

**体长：** 9～10 厘米

**生活习性：** 繁殖于高山森林中，迁徙时可见于低山次生林、城市绿化带、公园、果园及红树林等地。

Protection categories: 3V/LC

Status: Winter visitor

Body length: 9–10cm

Habitat and Habits: Breeding in high mountain forests, can be seen during migration in low mountain secondary forests, urban green areas, parks, orchards, and mangroves.

jù zuǐ liǔ yīng

# 巨嘴柳莺

# Radde's Warbler

㊫ *Phylloscopus schwarzi*

**保护等级：** 三有 / 无危

**居留类型：** 冬候鸟

**体长：** 12 ～ 14 厘米

**生活习性：** 常单独或成对活动于山脚平地的灌丛中，对水环境的要求不高。性隐秘，警惕性强，常跳跃，遇到危险时迅速逃离，觅食时靠近地面。

**Protection categories:** 3V / LC

**Status:** Winter visitor

**Body length:** 12–14cm

**Habitat and Habits:** Often solitary or in pairs, found in shrubs in the foothills. Has low requirements for water environments. Secretive and alert. Often jump, escape quickly when in danger. Forage close to the ground.

hè liǔ yīng
# 褐柳莺

# Dusky Warbler

㊫ *Phylloscopus fuscatus*

**保护等级：**三有 / 无危

**居留类型：**冬候鸟

**体长：**11 ～ 12 厘米

**生活习性：**繁殖生境为河流、沼泽周围的低矮灌丛及森林，迁徙或越冬时见于各类灌丛，偏好于近水的生境。

**Protection categories:** 3V/LC

**Status:** Winter visitor

**Body length:** 11–12cm

**Habitat and Habits:** Breeding habitats include low shrubs and forests around rivers and marshes. During migration or wintering, they can be found in various types of shrubs, preferably close to water.

miǎn liǔ yīng

# 冕柳莺

# Eastern Crowned Warbler

㊫ *Phylloscopus coronatus*

**保护等级：** 三有 / 无危

**居留类型：** 过境鸟

**体长：** 11 ~ 12 厘米

**生活习性：** 活动于各种树林及林缘，喜好大树木冠层。活泼好动，在枝叶间不停跳跃觅食小型昆虫。

**Protection categories:** 3V/LC

**Status:** Passing visitor

**Body length:** 11–12cm

**Habitat and Habits:** Inhabit forests and forest edges and prefer in crown layers of big trees; lively and active; jump among the branches and prey on small insects.

dàn jiǎo liǔ yīng

# 淡脚柳莺

# Pale-legged Warbler

㊫ *Phylloscopus tenellipes*

**保护等级：** 无危

**居留类型：** 过境鸟

**体长：** 10 ～ 11 厘米

**生活习性：** 常单独或成对活动于溪流附近的低矮林层或者灌丛中，有规律性的下压尾巴动作，迁徙时见于各类生境。主要捕食昆虫的幼虫。

**Protection categories:** LC

**Status:** Passing visitor

**Body length:** 10–11cm

**Habitat and Habits:** Often solitary or in pairs, actively in lower forest or shrub layers near streams. Exhibits rhythmic tail flicking. Found in various habitats during migration. Primarily feed on insect larvae.

jí běi liǔ yīng

# 极北柳莺

# Arctic Warbler

*Phylloscopus borealis*

**保护等级：** 三有 / 无危

**居留类型：** 过境鸟

**体长：** 12 ~ 13 厘米

**生活习性：** 常单独活动于茂密的树林中高层，也见于人工果园等环境，迁徙时对环境不甚挑剔。主要捕食昆虫的幼虫。

**Protection categories：** 3V / LC

**Status：** Passing visitor

**Body length：** 12–13cm

**Habitat and Habits：** Often solitary, inhabit the middle to upper layers of dense forests and can also be found in orchards. Not picky about the environment during migration; primarily feed on insect larvae.

huá nán guān wén liǔ yīng

# 华南冠纹柳莺

# Hartert's Leaf Warbler

㊫ *Phylloscopus goodsoni*

**保护等级：** 无危

**居留类型：** 过境鸟

**体长：** 10～12 厘米

**生活习性：** 多活动于树冠层，也会在灌丛觅食。常做两翼轮流抖动的动作。以昆虫为食。

Protection categories: LC

Status: Passing visitor

Body length: 10-12cm

Habitat and Habits: Mostly inhabit crown layers, and sometimes forage in the shrubs; often shake two wings in turn; feed on insects.

dōng fāng dà wěi yīng
# 东方大苇莺

# Oriental Reed Warbler

学 *Acrocephalus orientalis*

**保护等级：**无危

**居留类型：**冬候鸟

**体长：**17 ~ 19 厘米

**生活习性：**常单独或成对活动于湿地生态环境的有茂密的芦苇或高芒草地中高层。性活跃。觅食时常站立于高处，不甚惧人。

Protection categories: LC

Status: Winter visitor

Body length: 17–19cm

Habitat and Habits: Often solitary or in pairs, active in wetland habitats with dense reeds or tall grasses; active, standing on elevated positions while foraging, not afraid of people.

hēi méi wěi yīng

# 黑眉苇莺

# Black-browed Reed Warbler

*Acrocephalus bistrigiceps*

**保护等级：** 三有 / 无危

**居留类型：** 冬候鸟

**体长：** 13 ~ 14 厘米

**生活习性：** 常单独活动于湿地环境的芦苇及高芒草的中低层。性胆怯，遇到危险很快钻入密丛中。觅食时活跃，常跳动。

**Protection categories:** 3V/LC

**Status:** Winter visitor

**Body length:** 13–14cm

**Habitat and Habits:** Solitary and active in wetland habitats with reeds and tall grasses. Timid and quickly seeks refuge in dense vegetation when facing danger. Actively foraging, and often jumping.

gāo shān duǎn chì huáng yīng

# 高山短翅蝗莺

# Russet Bush Warbler

㊫ *Locustella mandelli*

**保护等级：** 无危

**居留类型：** 冬候鸟

**体长：** 13 ～ 14 厘米

**生活习性：** 隐匿于林缘及开阔而多灌丛山麓的密丛中，高可至海拔 2800 米。性胆小而隐秘，觅食或移动时在灌丛下安静地快速穿行。

**Protection categories：** LC

**Status：** Winter visitor

**Body length：** 13–14cm

**Habitat and Habits：** Hidden in forests and open foothill with dense shrubs of altitude up to 2800 meters. Timid and hidden; forage or move quickly and quietly in the shrubs.

zōng shàn wěi yīng

# 棕扇尾莺

## Zitting Cisticola

㊫ *Cisticola juncidis*

**保护等级：** 无危

**居留类型：** 冬候鸟

**体长：** 10 ～ 14 厘米

**生活习性：** 常单独活动于湿地环境中，如沼泽地、水稻田等，常站立于突出的低矮芒草上或稻杆上，可悬停，性活跃，飞行能力强，可垂直起飞。

Protection categories: LC

Status: Winter visitor

Body length: 10–14cm

Habitat and Habits: Solitary and active in wetland habitats such as marshes and rice fields. Often perch on prominent low grasses or rice stalks, capable of hovering. Active, strong flight abilities, can take off vertically.

huáng fù shān jiāo yīng

# 黄腹山鹪莺

# Yellow-bellied Prinia

*Prinia flaviventris*

**保护等级：** 无危

**居留类型：** 留鸟

**体长：** 12 ~ 14 厘米

**生活习性：** 活动于丘陵等林缘的灌木中，也喜欢浸水的草丛，常单独或成对活动，领域性较强。以昆虫为食。

**Protection categories:** LC

**Status:** Resident bird

**Body length:** 12–14cm

**Habitat and Habits:** Inhabit shrubs of forest edges in hill, and prefer grass submerged by water; often inhabit alone or in pairs with strong territory; feed on insects.

chún sè shān jiāo yīng

# 纯色山鹪莺

# Plain Prinia

㊫ *Prinia inornata*

**保护等级：** 无危

**居留类型：** 留鸟

**体长：** 11 ～ 15 厘米

**生活习性：** 常活动于高草丛或者茅草丰富的环境中，竹林、花园等地也可见。以昆虫为食。

Protection categories: LC

Status: Resident bird

Body length: 11–15cm

Habitat and Habits: Often inhabits environments with tall grass or reeds, and also be seen in bamboo and garden, etc.; feeds on insects.

chánɡ wěi fénɡ yè yīnɡ

# 长尾缝叶莺

# Common Tailorbird

㊫ *Orthotomus sutorius*

**保护等级：**无危

**居留类型：**留鸟

**体长：**10～14 厘米

**生活习性：**常单独或者成对活动，喜灌丛，也善爬树干，常将植物 1～2 片叶子缝起来当巢。以昆虫为食。

**Protection categories：** LC

**Status：** Resident bird

**Body length：** 10–14cm

**Habitat and Habits：** Often inhabit alone or in pairs, prefer shrubs, and also good at climbing trees; often sew one or two leaves together for a nest; feed on insects.

huà méi
# 画眉

保护等级：国二 / 无危
居留类型：留鸟
体长：21 ～ 24 厘米
生活习性：单独或成对或小群活动于茂密的林层低层或灌丛中，善鸣叫，声音婉转。在各种开阔的栖息地上觅食。

# Hwamei

㊫ *Garrulax canorus*

Protection categories：II / LC
Status：Resident bird
Body length：21–24cm
Habitat and Habits：Solitary or in pairs or small groups, active in dense forest and shrubs edges. Good at melodious calls. Forage on the ground in various open habitats.

hēi liǎn zào méi
# 黑脸噪鹛

# Masked Laughingthrush

㊫ *Pterorhinus perspicillatus*

**保护等级：** 三有 / 无危

**居留类型：** 留鸟

**体长：** 21 ～ 24 厘米

**生活习性：** 常集群活动于树木和灌丛中，一般不做长距离飞行，飞行姿态笨拙。以昆虫、果实等为食。

**Protection categories:** 3V/LC

**Status:** Resident bird

**Body length:** 21–24cm

**Habitat and Habits:** Often form flocks in trees and shrubs; normally fly in short distances with clumsy posture; feed on insects and fruits, etc.

bái jiá zào méi

# 白颊噪鹛

# White-browed Laughingthrush

㊫ *Pterorhinus sannio*

**保护等级：** 三有 / 无危

**居留类型：** 留鸟

**体长：** 22 ～ 24 厘米

**生活习性：** 常三五成群活动于各种环境下的灌丛中高层。性嘈杂，不甚惧人，有时也到林层底下翻动落叶寻找食物。

**Protection categories:** 3V/LC

**Status:** Resident bird

**Body length:** 22–24cm

**Habitat and Habits:** Gather in small groups and are active in the middle to upper layers of shrubby habitats in various environments. Noisy, not afraid of people. Sometimes forage on forest ground by turning leaves over to look for food.

lì jǐng fèng méi

# 栗颈凤鹛

# Chestnut-collared Yuhina

*Staphida torqueola*

**保护等级：**无危

**居留类型：**留鸟

**体长：**12～15 厘米

**生活习性：**常集群活动于密林下层，或者灌丛中高层。主要以单一种群组成“鸟浪”移动觅食，有时也混入其他小型鸟类共同组成“鸟浪”，常沿着公路边的水道觅食。

Protection categories: LC

Status: Resident bird

Body length: 12-15cm

Habitat and Habits: Gather in clusters in the lower layers of dense forests or the upper layers of shrubs, forming "bird waves" to forage. Sometimes mixed with other small birds to form "bird waves". Feed along waterways along the roadside.

àn lǜ xiù yǎn niǎo

# 暗绿绣眼鸟

**保护等级：**三有 / 无危

**居留类型：**留鸟

**体长：**10 ～ 11 厘米

**生活习性：**性活泼而喧闹，常成群于树顶觅食小型昆虫、小浆果及花蜜。常见于林地、林缘、公园及城镇。

# Swinhoe's White-eye

学 *Zosterops simplex*

Protection categories: 3V/LC

Status: Resident bird

Body length: 10–11cm

Habitat and Habits: Lively and noisy, often seen in groups foraging for small insects, berries, and nectar in treetops. Common in forests, forest edges, parks, and urban areas.

bā gē
# 八哥

# Crested Myna

㊫ *Acridotheres cristatellus*

**保护等级：**三有 / 无危

**居留类型：**留鸟

**体长：**23 ～ 28 厘米

**生活习性：**常集小群或大群活动于开阔的草坪、耕地等地，不甚怕人，天线塔及电线上也常见其身影。常在地面昂首阔步行走、觅食，属杂食性。

**Protection categories:** 3V/LC

**Status:** Resident bird

**Body length:** 23–28cm

**Habitat and Habits:** Often form small or large flocks in open grassland, and fields etc.; not afraid of people; often seen on power towers and wires; often strut on the ground for foraging; omnivorous.

jiā bā gē
# 家八哥

# Common Myna
㊫ *Acridotheres tristis*

**保护等级：** 三有 / 无危

**居留类型：** 留鸟

**体长：** 24 ～ 27 厘米

**生活习性：** 活动于农耕地、村庄、公园绿地或开阔地。喜群居，在地面取食，属杂食性。

**Protection categories:** 3V/LC

**Status:** Resident bird

**Body length:** 24–27cm

**Habitat and Habits:** Often inhabit paddy fields, villages, urban parks, or open areas; forage in flocks on the ground; omnivorous.

sī guāng liáng niǎo
# 丝光椋鸟

# Red–billed Starling

*Spodiopsar sericeus*

**保护等级：** 三有 / 无危

**居留类型：** 冬候鸟

**体长：** 18 ～ 23 厘米

**生活习性：** 常成对或集大群分布于有林的开阔地，觅食于高大乔木上层。冬季集大群活动，也取食于地面，性嘈杂而易被发现。

**Protection categories:** 3V/LC

**Status:** Winter visitor

**Body length:** 18–23cm

**Habitat and Habits:** Often occur in pairs or large flocks in open areas, foraging in the upper canopy of tall trees. In winter, forage in large groups, feed on the ground, noisy and easy to find.

huī liáng niǎo
# 灰椋鸟

# White-cheeked Starling

㊫ *Spodiopsar cineraceus*

**保护等级：** 三有 / 无危

**居留类型：** 留鸟

**体长：** 19 ～ 23 厘米

**生活习性：** 活动于矮草地、城市公园、农田开阔地。集群在地面行走，常与其他椋鸟混群。杂食性，主食昆虫，也吃果实、种子等。

**Protection categories:** 3V/LC

**Status:** Resident bird

**Body length:** 19-23cm

**Habitat and Habits:** Inhabit low grass, urban park, and open farmland; form flocks walking in the ground, and often mix with other starlings; omnivorous; mainly feed on insects, and also feed on fruits and seeds, etc.

hēi lǐng liáng niǎo

# 黑领椋鸟

# Black-collared Starling

㊫ *Gracupica nigricollis*

**保护等级：** 三有 / 无危

**居留类型：** 留鸟

**体长：** 27 ～ 31 厘米

**生活习性：** 常成对或小群活动于农田等开阔的地方，也会在城市开阔草地活动。性凶猛，会抢其他鸟的食物，属杂食性。

Protection categories: 3V/LC

Status: Resident bird

Body length: 27–31cm

Habitat and Habits: Often inhabit in pairs or form small flocks in fields and other open areas, and also in open urban grassland; fierce, and will robot other birds' food; omnivorous.

huī bèi liáng niǎo

# 灰背椋鸟

# White-shouldered Starling

㊫ *Sturnia sinensis*

**保护等级：** 三有 / 无危

**居留类型：** 冬候鸟

**体长：** 17 ～ 21 厘米

**生活习性：** 活动于湿地的防风林及城郊开阔林地，常集群活动，常与其他椋鸟、八哥混群活动。杂食性，以果实、花蜜、昆虫为食。

Protection categories: 3V/LC

Status: Winter visitor

Body length: 17–21cm

Habitat and Habits: Often form flocks in windbreaks of wetland and suburban open woodlands; often mix with other starlings; omnivorous; feed on fruits, nectar and insects.

hǔ bān dì dōng

# 虎斑地鸫

# White's Thrush

㊫ *Zoothera aurea*

**保护等级：** 三有 / 无危

**居留类型：** 冬候鸟

**体长：** 26 ～ 30 厘米

**生活习性：** 常单独活动于密林低层，也常出没于溪流附近的灌丛中。性隐秘，地面活动为主，警惕性强，惊动时可飞到附近的树枝上，几乎不到达树冠层。

**Protection categories:** 3V/LC

**Status:** Winter visitor

**Body length:** 26–30cm

**Habitat and Habits:** Solitary, inhabiting the understory of dense forests and often found near streams in thickets. Primarily forage on the ground, remaining alert and quickly flying to nearby branches. Barely reach to crown canopy.

huī bèi dōng

# 灰背鸫

# Grey-backed Thrush

㊫ *Turdus hortulorum*

**保护等级：**三有 / 无危

**居留类型：**冬候鸟

**体长：**18 ～ 23 厘米

**生活习性：**常单独活动于树林中高层，常隐身于树冠层下方，如南方的大叶榕等树高层，有时候也下地或者在高树下方老旧房子房顶翻动落叶觅食。

**Protection categories:** 3V/LC

**Status:** Winter visitor

**Body length:** 18–23cm

**Habitat and Habits:** Usually solitary, inhabiting the upper levels of the forest, often concealed beneath the canopy, such as in tall trees like the banyan. Sometimes go down on the ground or on the roof to turn the leaves over to look for food.

wū huī dōng

# 乌灰鸫

# Japanese Thrush

㊫ *Turdus cardis*

**保护等级：** 三有 / 无危

**居留类型：** 冬候鸟

**体长：** 18 ~ 23 厘米

**生活习性：** 习性像灰背鸫，但警惕性更强，常单独活动于树林中低层，也隐身于树冠层下方，如南方的大叶榕、细叶榕等树上。主要以无脊椎动物为食，如蚯蚓等。

**Protection categories：** 3V / LC

**Status：** Winter visitor

**Body length：** 18–23cm

**Habitat and Habits：** Vigilant, forage alone in the lower levels of forests, concealed beneath the canopy of trees like the banyan, feeding on invertebrates, like earthworms.

wū dōng

# 乌鸫

**保护等级：**三有 / 无危

**居留类型：**留鸟

**体长：**28 ～ 29 厘米

**生活习性：**生境多样、分布区最广的鸫类，繁殖季鸣唱婉转而多变且善效鸣，有“百舌鸟”之称。

# Chinese Blackbird

㊫ *Turdus mandarinus*

**Protection categories:** 3V/LC

**Status:** Resident bird

**Body length:** 28–29cm

**Habitat and Habits:** The most widespread thrush species and live in very varied habitats, known for their melodious and diverse songs during the breeding season.

bái méi dōng

# 白眉鸫

# Eyebrowed Thrush

㊫ *Turdus obscurus*

**保护等级：**无危

**居留类型：**冬候鸟

**体长：**20～24 厘米

**生活习性：**常活动于湿地边上的树林，在林中低层或地面上寻找食物。性胆怯，受惊时飞到附近树上。迁徙季节也与其他鸫类混小群觅食。

**Protection categories:** LC

**Status:** Winter visitor

**Body length:** 20–24cm

**Habitat and Habits:** Often found near wetland edges in forested areas, foraging in the lower layers or on the ground. Timid, they fly to nearby trees when startled. Feed in small groups with other thrushes during migration season.

què qú
# 鹊鸲

# Oriental Magpie Robin

*Copsychus saularis*

**保护等级：** 三有 / 无危

**居留类型：** 留鸟

**体长：** 19 ～ 22 厘米

**生活习性：** 常活动于人类活动密集之地，喜欢开阔草地、灌丛等，站立时尾常上翘。属杂食性。

**Protection categories:** 3V/LC

**Status:** Resident bird

**Body length:** 19–22cm

**Habitat and Habits:** Often seen in places near human habitation, and prefer open grassland and shrubs, etc.; often turn up the tail when standing; omnivorous.

bái fù lán wēng

# 白腹蓝鹟

# Blue-and-white Flycatcher

㊫ *Cyanoptila cyanomelana*

**保护等级：**无危

**居留类型：**过境鸟

**体长：**14 ～ 17 厘米

**生活习性：**常单独或成对活动于林相较好的林中层或树枝，极少到地面觅食，常常伫立于枝条上，发现目标后突然飞出捕捉猎物，然后回到原来的或附近的枝条。

Protection categories：LC

Status：Passing visitor

Body length：14–17cm

Habitat and Habits：Often seen solitary or in pairs in the middle layers or on branches of well-vegetated forests. Rarely forages on the ground.

wū wēng

# 乌鹟

# Dark-sided Flycatcher

㊫ *Muscicapa sibirica*

**保护等级：** 三有 / 无危

**居留类型：** 冬候鸟

**体长：** 12 ~ 14 厘米

**生活习性：** 栖息于中低海拔山地的针阔混交林和针叶林，喜阴暗的树林。觅食于植被中上层。

Protection categories: 3V/LC

Status: Winter visitor

Body length: 12–14cm

Habitat and Habits: Inhabits middle to low altitude mountainous areas, preferring mixed forests and coniferous forests. It favors shady and vegetation-rich environments, feeds in middle to upper level of vegetation.

huī wén wēng

# 灰纹鹟

## Grey-streaked Flycatcher

㊫ *Muscicapa griseisticta*

**保护等级：**三有 / 无危

**居留类型：**冬候鸟

**体长：**13 ~ 15 厘米

**生活习性：**常单独活动于林中层或冠层，少至地面。常停栖于突出枝头，飞出捕食飞虫后回到原地。

**Protection categories:** 3V/LC

**Status:** Winter visitor

**Body length:** 13–15cm

**Habitat and Habits:** Often inhabit alone in middle layers or crown layers of forests, and seldom to the ground; often settle on protruding branches, and fly out to prey on insects and then return to the spots.

tóng lán wēng

# 铜蓝鹟

**保护等级：**无危

**居留类型：**冬候鸟

**体长：**13 ～ 16 厘米

**生活习性：**常单独活动于林子中高层，或者灌木的高层，常停歇于树枝枝头或者其他相对显眼的地方。性大胆，不甚惧人。遇到猎物时蹿出捕捉空中飞虫。

# Verditer Flycatcher

㊫ *Eumyias thalassinus*

Protection categories: LC

Status: Winter visitor

Body length: 13-16cm

Habitat and Habits: Frequently active alone in the upper levels of forests or high branches of shrubs; it perches on branches or prominently; less fear to people, captures flying insects in air.

bĕi huī wēng

# 北灰鹟

## Asian Brown Flycatcher

*Muscicapa dauurica*

**保护等级：** 三有 / 无危

**居留类型：** 冬候鸟

**体长：** 12 ～ 14 厘米

**生活习性：** 常单独活动于林中层或冠层，少至地面。常停栖于突出枝头，飞出捕食飞虫后回到原地。

Protection categories：3V /LC

Status：Winter visitor

Body length：12–14cm

Habitat and Habits：Often inhabit alone in middle layers or crown layers of forests, and seldom to the ground; often settle on protruding branches, and fly out to prey on insects and then return to the spots.

hǎi nán lán xiān wēng

# 海南蓝仙鹟

# Hainan Blue Flycatcher

㊫ *Cyornis hainanus*

**保护等级：** 无危

**居留类型：** 留鸟

**体长：** 13 ～ 15 厘米

**生活习性：** 常单独或成对活动于密林中层，在林中行为活跃、动作矫健，觅食时常瞅准时机短距离结束战斗，完成捕捉任务。

Protection categories: LC

Status: Resident bird

Body length: 13–15cm

Habitat and Habits: Frequently active alone or in pairs in the middle layers of dense forests. Agile and energetic movement, swiftly seizing opportunities to capture prey at close range.

雄鸟 ♂

雌鸟 ♀

lán gē qú
# 蓝歌鸲

# Siberian Blue Robin

㊫ *Larvivora cyane*

**保护等级：** 三有 / 无危

**居留类型：** 过境鸟

**体长：** 12 ～ 14 厘米

**生活习性：** 常单独活动于阴暗湿润的林下地面，善于地面行走，站姿平稳，移动时常保持腹部后方的稳定性，遇到危险时抬头，然后迅速飞离。

**Protection categories:** 3V/LC

**Status:** Passing visitor

**Body length:** 12–14cm

**Habitat and Habits:** Often solitary, it forages on the damp forest floor, displaying a steady posture while walking and maintaining stability. Head up and fly away quickly when frightened.

hóng wěi gē qú

# 红尾歌鸲

# Rufous-tailed Robin

㊫ *Larvivora sibilans*

**保护等级：**三有 / 无危

**居留类型：**冬候鸟

**体长：**13 ～ 15 厘米

**生活习性：**常单独活动于阴暗潮湿的地面，尤其是略微有斜坡的地面。性警惕且胆怯，常常躲在灌丛地面活动。

**Protection categories:** 3V / LC

**Status:** Winter visitor

**Body length:** 13–15cm

**Habitat and Habits:** Often solitary, it prefers the damp and shady ground, especially areas with slight slopes. It is alert and timid, often hiding in the undergrowth.

hóng hóu gē qú

# 红喉歌鸲

**保护等级：** 国二 / 无危

**居留类型：** 冬候鸟

**体长：** 14 ～ 16 厘米

**生活习性：** 常单独活动于水边灌丛、草丛。善于在地面行走和觅食，主食昆虫，也吃植物嫩芽。

# Siberian Rubythroat

㊫ *Calliope calliope*

Protection categories: II/LC

Status: Winter visitor

Body length: 14–16cm

Habitat and Habits: Often inhabit alone in shrubs near water and grass. Walk and forage on the ground; mainly feed on insects, and also feed on new plant shoots.

hóng xié lán wěi qú
# 红胁蓝尾鸲

# Orange–flanked Bluetail

㊫ *Tarsiger cyanurus*

**保护等级：** 三有 / 无危

**居留类型：** 冬候鸟

**体长：** 12 ～ 14 厘米

**生活习性：** 常单独活动于林层底部或公路边上的坡地灌丛边地面，主要在地面活动，受惊时飞到附近树枝上，善于地面行走。

**Protection categories:** 3V/LC

**Status:** Winter visitor

**Body length:** 12–14cm

**Habitat and Habits:** Often solitary, it prefers the forest floor or slopes near roadside shrubs, primarily foraging on the ground but can quickly fly to nearby branches when frightened.

雄鸟♂

雌鸟♀

紫啸鸫

Blue Whistling Thrush

zǐ xiào dōng
# 紫啸鸫

# Blue Whistling Thrush

㊫ *Myophonus caeruleus*

**保护等级：** 三有 / 无危

**居留类型：** 留鸟

**体长：** 29 ～ 35 厘米

**生活习性：** 栖于临河流、溪流或密林中的多岩石露出处。地面取食，受惊时常边飞边发出类似口哨的尖哨声，停息时尾羽常散开且上下摆动。

**Protection categories:** 3V/LC

**Status:** Resident bird

**Body length:** 29–35cm

**Habitat and Habits:** Often found near rivers, streams, or rocky outcrops in dense forests. Forages on the ground. When frightened, it makes a sharp whistling call while flying away; spreads and swings its tail up and down when it stops.

bái méi jī wēng

# 白眉姬鹟

# Yellow-rumped Flycatcher

㊫ *Ficedula zanthopygia*

**保护等级：**三有 / 无危

**居留类型：**过境鸟

**体长：**12 ～ 14 厘米

**生活习性：**常单独活动于林子中高层，飞行能力强，发现飞行的昆虫时迅速出击，捉到猎物后迅速返回附近更高的枝条上站立。主要以飞行的昆虫为食。

**Protection categories:** 3V/LC

**Status:** Passing visitor

**Body length:** 12–14cm

**Habitat and Habits:** Often solitary, it inhabits the upper layers of the forest. With strong flying abilities, it swiftly attacks flying insects, and then returns to higher perches with captured prey.

雄鸟♂

雌鸟♀

huáng méi jī wēng
# 黄眉姬鹟

# Narcissus Flycatcher
㊫ *Ficedula narcissina*

**保护等级：**三有 / 无危
**居留类型：**过境鸟
**体长：**13 ～ 14 厘米
**生活习性：**常活动于开阔林地、公园，通常在树木中上层活动，空中捕捉飞虫后往往飞回原位。

**Protection categories:** 3V/LC
**Status:** Passing visitor
**Body length:** 13–14cm
**Habitat and Habits:** Often inhabit open woodlands and urban parks, and often act in the middle and top layers of forests; fly out to prey on insects and then return to the spots.

qú jī wēng
# 鸲姬鹟

# Mugimaki Flycatcher

㊫ *Ficedula mugimaki*

**保护等级：** 三有 / 无危

**居留类型：** 过境鸟

**体长：** 12 ～ 14 厘米

**生活习性：** 常单独活动于林层的树冠层，有时也到低矮灌丛中。性活跃，常于枝头上跳跃，飞行能力也强，可以在较短距离内捕捉到猎物。

**Protection categories：** 3V / LC

**Status：** Passing visitor

**Body length：** 12–14cm

**Habitat and Habits：** Often forage alone in the forest canopy, and sometimes in low shrubs. Active, hopping on branches, and strong flying abilities to capture prey within short distances.

雄鸟 ♂

雌鸟 ♀

hóng hóu jī wēng
# 红喉姬鹟

# Taiga Flycatcher

㊫ *Ficedula albicilla*

**保护等级：** 三有 / 无危

**居留类型：** 冬候鸟

**体长：** 12 ~ 14 厘米

**生活习性：** 常单独活动于水边开阔落叶树林及林缘。主食昆虫，也吃果实。

Protection categories: 3V/LC

Status: Winter visitor

Body length: 12–14cm

Habitat and Habits: Often inhabit alone in open deciduous forests and forest edges; mainly feed on insects and fruits.

zhě hóng wěi qú
# 赭红尾鸲

# Black Redstart

㊫ *Phonenicurus ochruros*

**保护等级：**无危

**居留类型：**过境鸟

**体长：**14 ~ 15 厘米

**生活习性：**常单独活动于开阔的荒野处，如多灌丛的草原，也常活动于长满草的建筑废料堆放地，也可见多草地的房顶。站姿比较平，常抖尾。

Protection categories：LC

Status：Passing visitor

Body length：14–15cm

Habitat and Habits：They often forage alone in open wilderness areas, such as shrubby grasslands, and frequently inhabit overgrown areas with debris or grassy rooftops. Stand flat and often shake the tails.

běi hóng wěi qú
# 北红尾鸲

# Daurian Redstart

㊫ *Phoenicurus auroreus*

**保护等级：** 三有 / 无危

**居留类型：** 冬候鸟

**体长：** 13 ~ 15 厘米

**生活习性：** 常单独活动于开阔疏林、田野或公园绿地。领域性强。常站立于突出物上，尾部不停抖动。主食昆虫与无脊椎动物，偶尔吃浆果及种子。

**Protection categories:** 3V/LC

**Status:** Winter visitor

**Body length:** 13–15cm

**Habitat and Habits:** Often inhabit alone in open forests, fields and parklands; strong territorial; often standing on the protruding objectives, and shake the tail all the time; mainly feed on insects and invertebrates, and some times feed on berries and weeds.

雄鸟 ♂

雌鸟 ♀

lán jī dōng
# 蓝矶鸫

# Blue Rock Thrush

㊫ *Monticola solitarius*

**保护等级：** 无危
**居留类型：** 冬候鸟
**体长：** 20 ～ 23 厘米
**生活习性：** 常立于岩石、峭壁、屋顶、枯树等突出物上，伺机俯冲地面捕捉昆虫。

**Protection categories：** LC
**Status：** Winter visitor
**Body length：** 20–23cm
**Habitat and Habits：** Often stand on rock, cliffs, roofs, dead trees and other protrusions, waiting for an opportunity to dive on the ground to catch insects.

华南亚种 *pandoo*

菲律宾亚种 *philippensis*

dōng yà shí jí

# 东亚石䳭

# Stejneger's Stonechat

㊫ *Saxicola stejnegeri*

**保护等级：**三有

**居留类型：**冬候鸟

**体长：**12 ～ 14 厘米

**生活习性：**常单独或成对活动于开阔地区的灌丛或林缘、农田等。主食昆虫，也吃蚯蚓、蜘蛛及种子。

**Protection categories:** 3V

**Status:** Winter visitor

**Body length:** 12–14cm

**Habitat and Habits:** Often inhabit alone or in pairs, in shrubs, open areas, forest edges, and fields, etc.; mainly feed on insects, but also eat earthworms, spiders and seeds.

雄鸟 ♂

雌鸟 ♀

hóng xiōng zhuó huā niǎo
# 红胸啄花鸟

# Fire-breasted Flowerpecker

学 *Dicaeum ignipectus*

**保护等级：** 无危

**居留类型：** 留鸟

**体长：** 7～9 厘米

**生活习性：** 常单独或集小群活动于适合的生境，如长有丰茂桑寄生植物的各类榕树中，主要活动于树冠层。秋季繁殖。

**Protection categories:** LC

**Status:** Resident bird

**Body length:** 7-9cm

**Habitat and Habits:** Often forage alone or in small groups in suitable habitats, such as banyan trees with abundant mistletoe, primarily in the canopy; breed in autumn.

chà wěi tài yáng niǎo

# 叉尾太阳鸟

# Fork-tailed Sunbird

㊫ *Aethopyga christinae*

**保护等级：** 三有 / 无危

**居留类型：** 留鸟

**体长：** 9 ～ 11 厘米

**生活习性：** 活动于森林、村庄、城市等地，常振翅悬停以微曲的喙和管状长舌吸取开花植物花中的花蜜，也吃小型昆虫和嫩芽。

**Protection categories:** 3V/LC

**Status:** Resident bird

**Body length:** 9–11cm

**Habitat and Habits:** Inhabit forests, villages, and urban areas, etc.; often draw nectar from flowering plants with slightly curved beak and long tubular tongue; also feed on insects and buds.

mú què
# 麻雀

# Eurasian Tree Sparrow
*Passer montanus*

**保护等级：**三有 / 无危
**居留类型：**留鸟
**体长：**12 ～ 15 厘米
**生活习性：**常集群活动，偏好人类居住环境。性嘈杂，筑巢于屋檐下的洞穴或树洞中，常到地面或草坪中觅食。以草籽、果实或小虫为食。

**Protection categories:** 3V / LC
**Status:** Resident bird
**Body length:** 12–15cm
**Habitat and Habits:** Often form flocks and prefer human habitation; noisy; nest in cavities under eaves or of the trees; often forage on the ground or lawns and feed on insects, seeds or fruits.

bái yāo wén niǎo

# 白腰文鸟

# White-rumped Munia

㊫ *Lonchura striata*

**保护等级：** 无危

**居留类型：** 留鸟

**体长：** 10 ～ 12 厘米

**生活习性：** 常见于低海拔的林缘、次生灌丛、农田及花园。性喧闹吵嚷，结小群生活。

Protection categories: LC

Status: Resident bird

Body length: 10–12cm

Habitat and Habits: Commonly found at low altitudes in forest edges, secondary shrubs, farmland, and gardens. Noisy and forgage in small flocks.

斑文鸟

Scaly-breasted Munia

bān wén niǎo
# 斑文鸟

**保护等级：** 无危
**居留类型：** 留鸟
**体长：** 10 ～ 12 厘米
**生活习性：** 常集大群活动，喜欢开阔的农田或者茅草区，会与白腰文鸟混群。喜啄食种子、果实。

# Scaly-breasted Munia

*Lonchura punctulata*

**Protection categories：** LC
**Status：** Resident bird
**Body length：** 10–12cm
**Habitat and Habits：** Often form large flocks, prefer open fields or reeds, and mix with White-rumped Munia; feed on seeds and berries.

shān jí líng

# 山鹡鸰

# Forest Wagtail

*Dendronanthus indicus*

**保护等级：** 三有 / 无危

**居留类型：** 过境鸟

**体长：** 16 ～ 18 厘米

**生活习性：** 单独或成对在开阔森林地面穿行。尾轻轻往两侧摆动，不似其他鹡鸰尾上下摆动。也停栖树上。

**Protection categories:** 3V /LC

**Status:** Passing visitor

**Body length:** 16–18cm

**Habitat and Habits:** Walk alone or in pairs on the ground in open forest, tail gently swinging side to side, unlike other wagtails. Also perch on trees.

huáng jí líng
# 黄鹡鸰

# Eastern Yellow Wagtail
㊫ *Motacilla tschutschensis*

**保护等级：** 三有 / 无危
**居留类型：** 冬候鸟
**体长：** 16 ～ 18 厘米
**生活习性：** 常集群活动于池塘湿地、沼泽、湖边、海边、水田及草地，常上下摆动尾羽。以昆虫为食。

**Protection categories:** 3V / LC
**Status:** Winter visitor
**Body length:** 16–18cm
**Habitat and Habits:** Often form flocks in ponds, marshes, lakes, coasts, paddy fields and grasslands; Often wag the tail up and down; feed on insects.

bái jí líng
# 白鹡鸰

# White Wagtail

㊫ *Motacilla alba*

**保护等级：** 三有 / 无危

**居留类型：** 留鸟

**体长：** 17 ～ 20 厘米

**生活习性：** 常活动于开阔的草地、湿地，也会在路上快速行走觅食，惊飞时几乎只会往前飞行，飞行姿势呈波浪状。食物以水边昆虫为主，也吃蜘蛛等无脊椎动物。

**Protection categories:** 3V/LC

**Status:** Resident bird

**Body length:** 17-20cm

**Habitat and Habits:** Often inhabit open grasslands and wetlands; forage by walking quickly and picking up items, and almost only fly forward when startled with wave-like flying posture; mainly feed on insects near water, but also eat spiders and other invertebrates.

huī jí líng
# 灰鹡鸰

保护等级：三有 / 无危
居留类型：冬候鸟
体长：16 ～ 18 厘米
生活习性：常单独活动于中低海拔山涧溪流、水域附近，上下摆动尾羽，静立时尾羽与地面平行。食物以水边昆虫为主，以小跑步方式或冲到空中捕食飞虫。

# Grey Wagtail
㊫ *Motacilla cinerea*

Protection categories：3V/LC
Status：Winter visitor
Body length：16–18cm
Habitat and Habits：Often inhabit alone around streams and other watery habitats at low and middle altitude; wag tail up and down; tail would be parallel to the ground when standing; mainly feed on aquatic insects, run and pick, or rush to the air to catch flying insects.

tián liù
# 田鹨

# Richard's Pipit

㊫ *Anthus richardi*

**保护等级：**三有 / 无危
**居留类型：**冬候鸟
**体长：**17 ～ 18 厘米
**生活习性：**常活动于农田、荒野，或开阔的矮草地。站姿挺直，波浪状飞行。捕食甲虫、蚂蚱等昆虫。

Protection categories：3V/LC
Status：Winter visitor
Body length：17–18cm
Habitat and Habits：Often inhabit fields or wildness, and low grass; upright standing posture, wave-shape flight pattern; prey on beetles, grasshoppers, other insects, and larvae.

shù liù
# 树鹨

# Olive-backed Pipit

㊫ *Anthus hodgsoni*

**保护等级：** 三有 / 无危

**居留类型：** 冬候鸟

**体长：** 15 ~ 17 厘米

**生活习性：** 偏好树林生境，常集小群活动于丘陵的林缘、果园、草地等，地面行走，尾羽常上下摆动，受惊时会飞到附近树上隐匿。以昆虫及草籽为食。

**Protection categories:** 3V/LC

**Status:** Winter visitor

**Body length:** 15–17cm

**Habitat and Habits:** Prefer forest environments, and often form small flocks in forest edges in hills, orchards, and grasslands, etc.; walk on the ground, wagging tails up and down; fly to surrounding trees and hidden when frightened; feed on insects and weeds.

bái méi wú
# 白眉鹀

# Tristram's Bunting

㊫ *Emberiza tristrami*

**保护等级：** 三有 / 无危

**居留类型：** 冬候鸟

**体长：** 13 ～ 16 厘米

**生活习性：** 常单独或小群活动于林下灌丛中。性隐秘，常于阴暗处活动，极少到开阔地。在林下地面或灌丛中取食草籽或小型昆虫。

**Protection categories：** 3V / LC

**Status：** Winter visitor

**Body length：** 13–16cm

**Habitat and Habits：** Often forage alone or in small groups within the undergrowth of forests. They prefer shady areas and rarely venture into open spaces. Feeding on seeds and small insects.

lì ěr wú

# 栗耳鹀

# Chestnut-eared Bunting

㊫ *Emberiza fucata*

**保护等级：** 三有 / 无危

**居留类型：** 冬候鸟

**体长：** 14 ~ 16 厘米

**生活习性：** 常单独或集小群活动于草丛低矮且稀疏的淡水湿地环境中，也会在村庄附近的灌丛中活动。主要取食草籽及小型昆虫。

Protection categories: 3V/LC

Status: Winter visitor

Body length: 14–16cm

Habitat and Habits: Forage alone or in small groups in grassy wetlands, also in shrubs near villages; feeding on grass seeds and small insects.

xiǎo wú
# 小鹀

# Little Bunting

㊫ *Emberiza pusilla*

**保护等级：** 三有 / 无危

**居留类型：** 冬候鸟

**体长：** 11 ～ 14 厘米

**生活习性：** 常与鹨类混群。常隐于浓密覆盖下和芦苇地。也见于农田周围的草丛和灌丛中。

**Protection categories:** 3V/LC

**Status:** Winter visitor

**Body length:** 11–14cm

**Habitat and Habits:** Often associate with pipits. Seek over in dense vegetation and reed beds, also found in grass and shrubs near farmland.

tián wú

# 田鹀

# Rustic Bunting

*Emberiza rustica*

**保护等级：**三有 / 易危

**居留类型：**冬候鸟

**体长：**13 ~ 15 厘米

**生活习性：**活动于平原至山麓的开阔杂草地、杂木林及苗圃。胆大不怕人，停栖时常竖起冠羽。常集小群活动，地面觅食草籽，偶吃昆虫。

**Protection categories:** 3V / VU

**Status:** Winter visitor

**Body length:** 13–15cm

**Habitat and Habits:** Inhabit open weedy fields, mixed forests and nurseries from the plain to the foothills; less afraid of people; erect frown feather when resting; forage on the ground for grass seeds and eat insects sometimes.

lì wú
# 栗鹀

# Chestnut Bunting

㊫ *Emberiza rutila*

**保护等级：** 三有 / 无危

**居留类型：** 冬候鸟

**体长：** 14 ～ 15 厘米

**生活习性：** 活动于低海拔山麓林缘、田间树木、稻田和灌丛草地，不甚怕人，接近时才飞。以草籽和嫩芽为食。

**Protection categories:** 3V/LC

**Status:** Winter visitor

**Body length:** 14–15cm

**Habitat and Habits:** Inhabit forest edges of low-altitude foothills, paddy fields, shrubs and grasslands; less afraid of people and fly away when people close to them; feed on weeds and buds.

雄鸟 ♂

雌鸟 ♀

huī tóu wú

# 灰头鹀

# Black-faced Bunting

㊫ *Emberiza spodocephala*

**保护等级：** 三有 / 无危

**居留类型：** 冬候鸟

**体长：** 13 ～ 16 厘米

**生活习性：** 活动于林缘开阔地、农田、荒地、苗圃等地，生性机警，稍有干扰即迅速飞入附近的灌丛中。以草本植物种子为食，也吃昆虫。

**Protection categories:** 3V /LC

**Status:** Winter visitor

**Body length:** 13–16cm

**Habitat and Habits:** Inhabit open forest edges, farmlands, and scrub; alert and flying to nearby shrubs when frightened; feed on seeds, but also eat insects.

# 观察手记
# Observation Note

# 2020 年五四青年节
# 南湾遇见心心念念的黄嘴白鹭

2013 年珠海市观鸟协会成立之初，郭会长收到一本香港观鸟会赠送的书籍《香港鸟类图鉴》（香港观鸟会，2013），封面的主角是一只漂亮的白色鹭鸟，橘黄色的嘴，后枕部飘着长而飘逸的羽冠，帅气内敛的气质立刻俘获了她的芳心。

它是谁？香港 500 多种各式各样的鸟，缘何选它做封面？经过一番了解，原来它名叫黄嘴白鹭，由于栖息地的丧失，全球仅有 3000 多只，属于全球易危的珍稀鸟类，2021 年升级为国家一级重点保护动物。在我国，黄嘴白鹭有 1000 多只，在辽东半岛、山东、浙江和福建的沿海岛屿繁殖，秋冬季节沿着海岸线向南方迁徙。

如此珍稀的鸟儿，郭会长在心里就惦记上了。珠海距离香港 30 海里左右，最近的外伶仃岛只距其 6 海里，珠海的海岸线绵延 600 多千米、大小岛屿 146 个，水道纵横、环境优美，湿地生境与香港类似，又同处于东亚—澳大利西亚候鸟迁飞区。这一带的湿地为各种候鸟提供了良好的栖息和觅食场所，既然黄嘴白鹭选择香港作为迁徙途中补充能量的重要驿站，那么它一定也会在珠海出现吧。

于是，郭会长带领一拨又一拨的小伙伴，跑遍了珠海的海滩滩涂、红树林湿地、田间地头，寻找黄嘴白鹭的踪迹。这一找，就是八年。她总是不厌其烦地叮嘱大家，不要对珠海随处可见的白鹭群视而不见，野外见到的每一只鸟都必须仔细观察，在它们当中很可能隐藏着“宝贝”。这八年，郭会长在遥远的大连庄河见过、在阳江阳西见过、在距离家门口最近的江门新会银湖湾也见过，却怎么也没在珠海见到，不知不觉中便将在珠海找到黄嘴白鹭拧成了一个心结。

2020 年 5 月 4 日，恰逢五四青年节，珠海市观鸟协会生态调查队青年调查员小鱼儿、Jay、林碧珊在横琴岛开展野外调查，途经南湾一片开阔的滩涂，远远看到一小群白色鹭鸟正在悠闲地享受阳光沙滩，或专心觅食，或戏水洗浴，或打情骂俏，好一派热闹和谐的景象。

突然，迎面飞来一只白色鹭鸟落在距离调查队 50 米左右的有水滩涂，体形大小和白鹭相仿。小鱼儿通过双筒望远镜观察，该白色鹭鸟一落下就开始觅食，时而奔跑展开翅膀，强烈的阳光照射，虽有上升热气干扰，头上有如“辫子”的白色饰羽随风飘扬，显得风度翩翩。它一个侧身摆头动作，闪现了黄色的喙，这一抹黄色令人眼前一亮，双手紧握双筒望远镜，不敢眨眼持续观察，摆头回来那刻又看到黄色的喙。这难道是我们期待已久的黄嘴白鹭？小鱼儿激动地说：“快拍！那只很可能是黄嘴白鹭！！”Jay眼疾手快，立马拍下来回放，确定它就是黄嘴白鹭。

得到消息的郭会长激动不已，立刻放下手中所有的事务，马不停蹄地赶到现场，可惜它已经羞羞答答地不知藏身何处了……

经查阅中国观鸟记录中心，这是继 10 年前（2010 年 4 月）在珠海三灶之后，珠海市第二笔黄嘴白鹭观测记录，也是珠海地区第一笔野外清晰的影像记录。同时，黄嘴白鹭更新了横琴岛鸟类记录。

在郭会长的倡议下，由珠海市观鸟协会发起，联合深圳市观鸟协会、广州市自然观察协会、湛江市爱鸟协会，共同承接了广东省林业局的委托，开展了为期一年的广东省七市（珠海、深圳、广州、江门、湛江、中山、阳江）黄嘴白鹭等珍稀水鸟同步调查项目。

项目参与机构积极开展野生动物（候鸟）野外巡护、协助保护执法、提供信息动向、公众教育、社区共管等工作，初步摸清了七个城市黄嘴白鹭等珍稀水鸟的分布情况，同时，开展候鸟迁飞路线上的巡护和监测工作，用科学和系统的方法记录广东重要湿地候鸟迁飞的数据，为利用粤港澳大湾区内滨海地带的联通性，开展候鸟迁飞保护、水鸟栖息生境修复、科普宣教建设和生态旅游建设提供可信的数据支持，携手助力提升生态廊道节点质量，共同将粤港澳大湾区打造成“候鸟湾区”。

# In 2020, Meet a Chinese Egret in Nanwan on Youth Day

At the beginning of the establishment of Zhuhai Bird Watching Society in 2013, President Guo received a book, "A PHOTOGRAPHIC GUIDE TO THE BIRDS OF HONG KONG" (Hong Kong Bird Watching Society, 2013), the main character on the cover was a beautiful white egret with an orange beak, and a long elegant crest floating on the backside of the head. The handsome and restrained temperament immediately caught her eyes.

Who is it? With more than 500 different kinds of birds in Hong Kong,why was it chosen for the cover of a book? It turned out that it was originally called Chinese Egret. Due to the loss of its habitat, it is estimated that there are only around 3000 left in the world, and it is a rare and vulnerable bird in the world. In 2021, it was upgraded to a national first class key protected animal. In China, there are over 1000 Chinese Egrets, which breed in the coastal islands of Liaodong Peninsula, Shandong, Zhejiang, and Fujian, and migrate south along the coastline in autumn and winter.

As such a rare bird, President Guo kept thinking about it. Zhuhai is approximately 30 nautical miles away from Hong Kong, with the nearest Wailingding Island just 6 nautical miles away. Zhuhai's coastline stretches over 600 kilometers, with 146 islands of various sizes. The waterways crisscross the area, creating a beautiful environment. The wetland habitat is similar to that of Hong Kong, and it is also located in the migration area of East Asian-Australasian Flyway. The wetlands in this area provide a good habitat and foraging place for various migratory birds. If the Chinese Egret had chosen Hong Kong as a "stopover" sites to refuel during migration, it would highly likely appear in Zhuhai as well.

Therefore, President Guo led groups of investigators to search for the tracks of Chinese Egret in Zhuhai, exploring everywhere including beaches, mangrove wetlands, and fields. The search lasted for eight years, during which she consistently urged everyone not to turn a blind eye to the egrets everywhere in Zhuhai, encouraging them to carefully look and observe at every bird they saw in the wild, as there might be a "treasure" hidden among them. In the past eight years, President Guo has seen Chinese Egrets in Zhuanghe, Dalian, Yangxi, Yangjiang, and Yinhuwan in Jiangmen Xinhui, the nearest place to home, but never in Zhuhai. Therefore, finding the Chinese Egret in

Zhuhai had become her primary goal and task.

On May 4, 2020, the Youth Day, three investigators of the investigation team, Little Fish, Jay, and Lin Bishan, conducted a field investigation on Hengqin Island. After passing through an open beach in Nanwan, they saw a small group of egrets leisurely enjoying the sunny beach, concentrating on foraging, bathing in the water, or playing with each other, creating a harmonious scenery.

Suddenly, a white egret flew in front of them and landed on the mudflat about 50 meters away from the investigation team. Its size was similar to that of an egret. Little Fish observed it through binoculars. The white egret was in search of fish as soon as it landed, and it ran around and spread its wings. Despite the strong sunlight and the rising heat, the white feathers on its head resembled "pigtails" fluttering in the wind, giving it an elegant appearance. It swayed its head sideways and flashed its yellow beak. Little Fish's eyes lit up upon seeing the yellow beak, and holding the binoculars with both hands, he watched attentively without blinking. The moment it shook its head back, he caught sight of the yellow beak again. Excitedly, Little Fish exclaimed, "Take a picture! That may be a Chinese Egret!" Jay quickly noticed it and immediately recorded it for playback, confirming that it was a Chinese Egret.

President Guo, upon receiving the news, was very excited. She immediately dropped all her affairs and rushed to the scene. Unfortunately, the Chinese Egret had already shyly hidden somewhere...

After consulting the China Bird Watching Record Center, this was the second observation record of Chinese Egret in Zhuhai, and the first clear field image record in Zhuhai since the observation in Sanzao ten years ago, in April 2010. At the same time, the Chinese Egret had set a new record of bird for Hengqin Island.

Upon the initiative of President Guo, Zhuhai Bird Watching Society, together with Shenzhen Bird Watching Society, Guangzhou Nature Association, and Zhanjiang Bird Watching Society, jointly undertook the commission of the Forestry Administration of Guangdong Province and conducted a one-year synchronized investigation project on rare waterfowls, including the Chinese Egret, in seven cities of Guangdong Province (Zhuhai, Shenzhen, Guangzhou, Jiangmen, Zhanjiang, Zhongshan, Yangjiang).

The participating agencies of the project actively conducted wild animal (migratory bird) field investigation, assisted in protection and law enforcement, provided information trends, public education, and community management. They preliminarily identified the distribution of rare waterfowl, such as Chinese Egrets, in seven cities, carried out investigation and monitoring on migratory bird migration routes, and recorded the migration data of migratory birds in important wetlands in

Guangdong with a scientific and systematic approach. It would provide credible data support to carry out migratory birds protection, waterfowl habitat restoration, science popularization, education construction, and ecological tourism construction. These data can also provide useful information about where migratory birds would stop for food and rest along the coastline of the Guangdong–Hong Kong–Macao Greater Bay Area, which would help those cities in the Great Bay Area working together to help improve the quality of ecological corridor nodes and jointly create the Guangdong–Hong Kong–Macao Greater Bay Area into a "Migratory Birds Bay Area".

# 2020年初冬望眉角数出红头潜鸭

每年11月进入深秋初冬，北方已是落叶纷飞、草木枯荣、雁鸭南飞，地处亚热带的珠海却逐渐迎来了“青芜不尽天连水，野鸭飞来成画图”的壮丽场景。

横琴岛的望眉角、二井湾、天沐河、南湾、石栏洲等地是观测雁鸭类的绝佳场所，特别是望眉角，融浅海水域、淤泥质滩涂、鱼虾水产养殖场、红树林、灌丛、常绿阔叶林等为一体，生境类型多样化且人为干扰少，适于鸟类，尤其是鸭科鸟类在此栖息和觅食。

鸭科鸟类喜欢集群，为了获得准确的种群、密度等调查数据，“数鸭”是珠海市观鸟协会生态调查队的基本功。郭会长带领调查队开展野外调查，每逢遇到野鸭成大群集结的时候，总是把这句口头禅挂在嘴边：“架上单筒望远镜，把鸭子从头到尾仔细地数一遍。”

号称目光犀利、眼疾手快的心脏介入专家军长，每年完成复杂的心脏介入手术几百台，与死神赛跑挽救患者生命，是一位令人非常敬佩和尊重的医学专家。他还义务作为调查队的高级医学顾问及生态志愿者，业余时间喜欢与调查队一起外出拍鸟，凭着职业的敏感，他抓拍飞鸟的速度和水平堪称一流。可别看他在手术台上使用各种导管镇定自若，操作自如，但凡遇到操作单筒望远镜“数鸭”的任务却有点儿发怵。

“术业有专攻嘛……”有时候数着数着，鸭子扑棱扑棱飞跑了，军长一时间数不清楚了，总会自我开脱一下，“做手术论精准，你们肯定比不上我，可是这野外数鸭的活儿，还是你们更专业！”

“数鸭”不是数数那么简单，还真的是一门技术活儿。首先，要选择视角好且不容易被鸟儿发现的地方放置单筒望远镜。其次，熟练操作高倍率的单筒望远镜，精确地找到鸟也是需要经常练习的，初学者通常会一通乱扫，对不上焦；熟手也会因为阳光炙热的天气产生的晕光或者阴雨雾霾的天气造成的能见度差，而能够辨识目标的能力不同。最后，必须具有持之以恒的耐心，鸟儿是灵动的，会飞会跑，可能瞬间就“逃离”单筒望远镜的视野。

2020年11月17日，调查队顶着蒙蒙细雨来到横琴岛进行样线调研，先是路过几个虾塘，见到绿翅鸭、白眉鸭、琵嘴鸭、赤颈鸭、针尾鸭混群，捕捉食物，追逐嬉戏，好不惬意，令湿漉漉的小伙伴们顿时心情大好，不辞辛劳地继续前往望眉角。

在距离岸边 120 米左右的海面上，调查员小鱼儿发现有一大群黑魆魆的物体随着波浪上下晃动着，他立刻熟练地架起单筒望远镜，通过观察，发现大约有上千只的凤头潜鸭，这可够“数”的啦！

阴雨绵绵，海面灰蒙蒙的，能见度极差，潜鸭们有的把头蜷缩在翅膀下昏昏睡觉，有的顽皮得很，一会儿潜进水里觅食，一会儿又从另外一处冒出头来，打乱小鱼儿“数鸭”的节奏。他数了一遍又一遍，持续观察了两个多钟头。

突然，一只体形稍大一点、与众不同的潜鸭引起了小鱼儿的注意，他赶紧拿出手机，接上单筒望远镜准备拍摄，而这只潜鸭更是调皮，跟他玩起了捉迷藏，时而潜水，时而躲到其他潜鸭身后……小鱼儿却是极有耐心，不依不饶，终于追踪拍摄到清晰的照片，确认了它是混在凤头潜鸭大群中的一只红头潜鸭雄鸟，而且旁边还有两只红头潜鸭雌鸟相伴，其乐融融，颇为惬意，为横琴岛收获了一个鸟类新记录。

# In 2020, Three Common Pochards are counted at Wangmeijiao in early winter

Every November, it is getting colder and colder. In northern China, leaves are falling; plants are withering, and geese and ducks will migrate to the south. Meanwhile, in southern China, the subtropical city–Zhuhai presents a magnificent scene quite different: the endless green reeds are thriving and groups of geese and ducks migrate to Zhuhai.

Wangmeijiao, Erjingwan, Tianmu River, Nanwan, and Shilanzhou on Hengqin Island are excellent places for observing geese and ducks, especially Wangmeijiao, which integrates shallow waters, silty mudflats, fish and shrimp aquaculture farms, mangroves, bushes, and evergreen broad–leaved forests. With diverse habitats and less human interference, it is suitable for birds, especially Anatidae birds, to perch and forage.

Anatidae birds like to gather in clusters, so "counting ducks" is the basic skill of the project investigation team to obtain accurate population and density data. During President Guo led the investigation team in the field, whenever she encountered large flocks of ducks, she always says: "Set up monoculars, and count the ducks carefully from beginning to end."

One of our members, nicknamed "Military Commander", is a cardiovascular intervention expert with sharp eyes and quick hands, and he has completed hundreds of complex heart intervention operations each year, saving a lot of lives. Highly respected and admired, he also serves as a senior medical consultant and ecological volunteer for the investigation team. In his spare time, he enjoys accompanying the team on bird photography expeditions. With his professional sensitivity, he possesses exceptional speed and skill in capturing candid photographs of birds. Despite his calm and professional demeanor while using various catheters during surgeries, he feels a little nervous when tasked with counting ducks using a monocular.

"Each profession has its own expertise..." Occasionally, when the ducks flutter and fly away while being counted, The Military Commander is temporarily unable to count them. In such moments, he often jokingly excuses himself, saying, "You guys may not be as precise as me during surgery, but you are certainly more professional in counting ducks!"

"Counting ducks" is not simply counting numbers but a technical work. First of all, you should choose a place with a good view that is not easy for birds to discover, and set up your monocular. Secondly, mastering the operation of high-magnification monocular and finding birds accurately require a lot of practices. Beginners usually scan all the time without focusing, and skilled hands also have difficult in identifying targets due to the halation caused by a hot sunny day, or poor visibility caused by a rainy hazy day. Moreover, persistent patience is necessary. Birds are agile, capable of flying and running, and may escape from the vision of monocular instantly.

On November 17, 2020, a drizzly day, the investigation team came to Hengqin Island to investigate on the line transect. First, they passed several shrimp ponds and saw a group of Eurasian Teal, Garganey, Northern Shoveler, Eurasian Wigeon, and Northern Pintail, foraging, chasing, and playing. This sight delighted the team despite the wet conditions, and they continued their journey to Wangmeijiao with renewed vigor.

About 120 meters away from the shore, investigator Little Fish found a large group of dark objects shaking up and down with the waves. He immediately set up his monocular skillfully. Through observation, there were more than a thousand of ducks and geese in this large group, which would take a long time to count!

It was raining continuously, the sea was grey, and the visibility was very poor. Some of the ducks curled their heads under wings and slept in a daze. Some were very naughty. Sometimes they went into the water to forage, and then emerged from another place, disrupting Little Fish rhythm of "counting ducks". He counted them over and over, and continued to observe for more than two hours.

Suddenly, a slightly larger diving duck attracted Little Fish attention. He quickly took out his mobile phone, connected to the monocular, and prepared to take a picture of this special diving duck. This diving duck was even more naughtier. It played hide and seek with him, sometimes diving, sometimes hiding behind other diving ducks.

With its red head contrasting sharply against its white back, Little Fish wondered if it was a Common Pochard. Nonetheless, he was very patient and persevered. Finally, he tracked and took clear pictures, confirming that it was a male Common Pochard, with two female Common Pochards nearby. A new bird record was harvested for Hengqin Island.

# 2020 年望天台水库“听声辨鸟”识蝗莺

对于观鸟爱好者而言，各种莺是他们又爱又恨的小家伙。说到“爱”，莺是雀形目柳莺科、蝗莺科、苇莺科等一大群小型鸟类的统称，在我国计有 100 多种。它们大多数在 10 厘米左右，颜色以灰褐色系和黄绿色系为主，身材玲珑，娇小可爱。再说“恨”，大部分莺喜欢在树冠和灌丛中活动，而且生性活泼，跳来跳去很少停歇，观察起来比较困难，加上它们的特征太过于相似，所以要想一眼看过去就能准确辨识更是难上加难。

好在珠海市观鸟协会生态调查队的队员们拥有“听声辨鸟”的绝技，他们在借助望远镜观鸟的同时，一直在用耳朵“观”鸟。声音通常是探测鸟类存在的最佳方式，尤其是对于特征不明显的莺系列的辨识主要靠声音，在那些稠密的树枝缝隙中，仅靠眼睛去看根本不行。首先要研究鸟鸣声，精准判断鸟的类型，再播放鸟鸣声引其现身，拍摄到清晰的鸟类图片，更大程度地增加调查的可信度。

2020 年 11 月 19 日，调查队在横琴岛望天台水库边的人工次生林开展野外调研时，发现一只莺在眼前的树上一晃而过，迅速躲进树丛中消失不见了。待要离开，它却在藏身处 rink-tink-tink 兴奋地鸣叫，调查员小鱼儿循声追赶过去，鸟儿又扑棱棱飞到旁边的树丛中躲起来，好像在跟他捉迷藏一样。

一来二去，惹得小鱼儿有些恼火，干脆不去追赶，停下来仔细倾听鸟儿的叫声，然后拿出手机找到一个鸟鸣文件播放起来。这一下轮到鸟儿迷惑了，明明自己形单影只从千里迢迢的北方历尽千辛万苦迁徙到这里越冬，难道还有同族也飞到了这里跟自己抢地盘吗？趁着它气势汹汹地从藏身处一跃而起，跳上一根裸露的树枝寻找“入侵者”算账的时候，小鱼儿眼疾手快，按动快门，迅速地拍下了它的倩影。放大照片仔细观察，这只鸟正是小鱼儿之前预判的高山短翅蝗莺！

凭着“听声辨鸟”的方法，调查队陆续发现了高山短翅蝗莺、华南冠纹柳莺、栗鹀等横琴岛新记录，它们也是珠海市鸟类新记录。其实，或许它们早就造访过这里，只是由于科学性、系统性的生态调查不足，而普通人的观察能力有限，才没有被登记在册的吧。

# In 2020, Identifying Warbler by Bird Sound at Wangtiantai Reservoir

For bird watching enthusiasts, various species of warblers are both loved and hated little creatures. Speaking of "love", warblers are a collective term for a large group of small birds belonging to the Sylviidae, Locustellidae, and Acrocephalidae families within the Passeriformes order. There are over 100 species of warblers in China. Their body length is generally around 10 cm, with colors primarily ranging from gray–brown to yellow–green. They have petite and adorable figures. Now, speaking of "hate", most warblers prefer to move around in the canopies of trees and shrubs. They are naturally lively, constantly hopping around with little rest, which makes observation quite challenging. Moreover, their similar appearances make accurate identification even more difficult at a glance.

Fortunately, the members of the Bird Watching Society's Ecological Survey Team in Zhuhai have the remarkable skill of "identifying birds by their calls". While observing birds with the help of binoculars, they rely on their ears to "observe" birds. Sound is often the best way to detect the presence of birds, especially for identifying species in the yuhina family which physical features are not distinct. In dense tangles of branches, relying solely on eyesight is inadequate. First, it is necessary to study bird songs to accurately identify the type of bird, and play the bird songs to lure them into appearing. Then capturing clear images of birds by camera can greatly increase the credibility of the identification of bird species.

On November 19, 2020, during a field survey in the secondary forest near the Hengqin Island's Wangtiantai Reservoir, the survey team spotted a warbler. It quickly darted across a tree right in front of them and disappeared into the foliage. As they were about to leave, the bird suddenly chirped excitedly with a "rink–tink–tink" call from its hiding spot. Investigator Little Fish followed the sound and chased after it, but the bird fluttered away and hid in the nearby thicket, as if playing a game of hide–and–seek with him.

As the back–and–forth continued, it started to annoy Little Fish. He decided to stop chasing it ,while listen carefully to the bird's calls. He took out his phone and played a bird call record. This confused the bird since it migrated alone all the way from the far north, overcoming numerous hardships to live through the winter here.

Did another member of its species also arrive here and invade its territory? Seizing the moment when the bird vigorously leaped from its hiding spot and jumped onto a bare branch to search for the "intruder", Little Fish swiftly pressed the shutter button and quickly captured its graceful figure. Upon zooming in and closely examining the photo, it was undoubtedly the Russet Bush Warbler that Little Fish had previously anticipated!

Using the method of "identifying birds by their calls", the survey team gradually discovered new records in Hengqin Island, such as the Russet Bush Warbler, Hartert's Leaf Warbler, and Chestnut Bunting. These findings also represented new records for bird species in Zhuhai City. In fact, it is possible that these birds had visited the area long before, but due to insufficient scientific and systematic ecological surveys and the limited observation abilities of ordinary people, they were not officially recorded until now.

# 2020 年神秘的“海豚湾”观鸻鹬

横琴岛上隐藏着一处未被开发的“处女地”，驱车来到横琴西南角，已经没有了前进的道路，海岸边矗立着一堆大小不一的礁石。珠海市观鸟协会生态调查队队员八哥领着小伙伴们驾轻就熟、小心翼翼地穿过礁石堆，眼前豁然开朗，一望无际湛蓝湛蓝的大海、雪白雪白的沙滩，立刻就有了置身于度假胜地的感觉。据当地的渔民说，在远处布满奇石附近的海域，曾经出现过几十只白海豚，民间称之为“海豚湾”。

如此宁静而神秘的好地方，不仅能够让人由衷地欢喜，更是鸻鹬类水鸟的“天堂”。为了深入了解“海豚湾”的生物多样性海域，八哥已经连续数年到此地观测水鸟。夏季的时候，曾经观测到数百只金眶鸻在此繁殖；春、秋、冬三季，迁徙或者过境的鸻鹬类水鸟更是喜欢在此栖息、觅食。由于羽色接近、不同季节羽色变化大、喜欢混群活动等特性，给鸻鹬类水鸟的辨识增加了不少的挑战性。

八哥是珠海的小学老师，也是郭会长一手从“菜鸟”培养起来的观鸟达人。郭会长称那些认识不到 100 种鸟的人为“菜鸟”，然而她却从来都不歧视“菜鸟”。“谁还没有一个从菜鸟成长起来的过程呢？”她总是鼓励热爱鸟儿的人，尤其重视培养八哥这样的中小学老师，乃至大学老师：“要把他们培养出来，再借助他们的力量影响孩子们，让孩子们从小就开始喜欢观鸟，然后爱鸟、护鸟，进而热爱我们的生存环境，促进自然生态保护可持续健康发展。”

如今的八哥，已经成为能够熟练分辨鸻鹬类水鸟的达人了，他会教孩子们首先通过观察行为分辨鸻和鹬。鸻类因为是依靠大眼睛来发现食物，它们会站立在原地寻找食物，待发现后快速奔跑过去啄食，因此，它们的行为模式是“站立—奔跑—啄食—站立—奔跑—啄食”；而鹬类大部分是用嘴插到泥土中依靠触觉感知食物，它们经常是在同一个地方不断地用嘴在泥地中探索食物，也可以边走边探索，因此，它们的行为模式是“走—探—吃—走—探—吃”。

他还教会孩子通过嘴的形状和长短来区别鹬类。从嘴的形状来举例：身材高挑有细直嘴的是黑翅长脚鹬，同样高挑但有上翘嘴的是反嘴鹬。从嘴的长短来辨识更是技术活，一般用鸟嘴的长度与头的长度来进行比较，拿我国分布的四种杓鹬进行比较：小杓鹬的嘴长约为头长的 1.5 倍，中杓鹬的嘴长约为头长的 2 倍，而大杓鹬和白腰杓鹬的嘴长均达到头长的 3 倍或以上。在八哥的影响下，他所在的学校培养出一批又一批的观鸟爱好者，这些孩子逐渐长大后，又带动了身边的家长、同学、小伙伴，给他们传递鸟类知识、生态环保理念，共同呵护珠海美丽的生态家园。

2020 年 11 月 21 日，阳光明媚，风和日丽，调查队再次来到“海豚湾”，置身于蓝天碧海的美景，枯燥的调查工作也成了一种享受。正赶上退潮时分，远处的滩涂有一大群灰麻麻的物体在动来动去，通过双筒望远镜，初步判断是鸻鹬类的鸟儿。调查员小崔猫下身子，迅速朝着它们的方向赶了过去，架好单筒望远镜仔细观察起来。

这次的收获真是颇为丰盛，有体背红褐色、白色头和腹部与胸前醒目黑色胸兜明显对比的翻石鹬，有头顶红色、胸前具黑色细纵纹、脚黄色的长趾滨鹬，还有头顶红色、胸前自带黑色 V 形斑的尖尾滨鹬，它们都是横琴岛的新记录。

# In 2020, Watching Red-necked Plover and Sandpiper in the Mysterious "Dolphin Bay"

A mysterious and undeveloped "virgin land" hides in Hengqin Island. What is it? We drove to the southwest corner of Hengqin, and there is no road ahead. Along the coastline, there stand a pile of rocks of various sizes. The Zhuhai Birdwatching Society's ecological survey team, led by "Bage", one of the team members, skillfully and cautiously navigate through the rocks, and suddenly, a vast expanse of the deep blue sea and pristine white sandy beach caught our sight. They immediately feel as if they have entered a holiday resort. According to local fishermen, dozens of Chinese White Dolphins have been spotted in the coastal area with rocks of various size and shape, where is known locally as "Dolphin Bay".

Such a peaceful and mysterious place not only brings genuine joy to people but also serves as a "paradise" for avocet waterbirds. In order to gain a deeper understanding of the biodiversity in the "Dolphin Bay" marine area, "Bage" has been observing shorebirds here for several years. During the summer, hundreds of Little Ringed Plovers have been observed breeding in this area. In spring, autumn, and winter, various migratory shorebirds prefer to rest and forage here. It brings a considerable challenge to identify shorebirds due to the similar color of their plumage of different species, significant variations in feather colors of a certain species in different seasons, and the tendency of several species mixed in flocks.

"Bage" is a primary school teacher in Zhuhai and also a bird-watching enthusiast who was nurtured by President Guo from being a "novice". President Guo refers to those who know less than 100 bird species as "novices", but she never looks down upon them. "Who hasn't gone through the process of growing from a novice?" she always encourages bird lovers, especially valuing the development of teachers like "Bage", including primary and even university teachers. "We need to cultivate the teachers and let them to influence children, so that children can enjoy bird watching, love and protect birds from a young age; leading to an affection for our living environment, ultimately promoting sustainable and healthy development in natural ecological conservation."

Today, "Bage" has become an expert at distinguishing Shorebirds and can skillfully teach children to differentiate between plovers and sandpipers based on their behavior. Plovers rely on their large eyes to detect food. They stand in one place, searching for food, and once they spot it, they quickly run over and peck at it. Their behavioral pattern is: "Stand – Run – Peck – Stand – Run – Peck." On the other hand, sandpipers mostly use their bills to probe into the mud, relying on their sense of touch to find food. They often stay in the same spot, continuously probing the mud with their bills to explore for food, or they can explore while walking. Their behavioral pattern is: "Walk – Probe – Eat – Walk – Probe – Eat." "Bage" teaches children to observe these behavioral patterns as the first step in distinguishing between avocets and sandpipers.

He also teaches children to differentiate sandpipers based on the shape and length of their bills. As an example of bill shape, he explains that the Black-winged Stilt is slender with a straight bill, while the Pied Avocet has a similarly slender body but a slightly upturned bill. Distinguishing sandpipers based on bill length is even more challengeable and requires technical skills. Typically, the length of the bird's bill is compared to the length of its head. For example, among the four species of Curlews found in China, the bill of the Eskimo Curlew is about 1.5 times the length of its head, the bill of the Whimbrel is about 2 times the length of its head, and both the Far Eastern Curlew and the Eurasian Curlew have bills that are 3 times or more the length of their heads. Under "Bage" 's influence, his school has nurtured a continuous stream of bird-watching enthusiasts. As these children grow up, they in turn inspire their parents, classmates, and friends, spreading bird-related knowledge and ecological conservation concepts. Together, they contribute to protect the beautiful ecology of Zhuhai.

It was a sunny and pleasant day on November 21, 2020, and the eco-investigation team arrived at "Dolphin Bay" once again. It was such an pleasure enjoyment for the tedious eco-survey work when the team members immersed in the beautiful scenery of blue skies and clear seas. Coinciding with the ebbing tide, a large group of grayish objects was moving in the distant mudflats. Through binoculars, it was preliminarily determined that those would be shorebirds. Investigator Xiao Cui crouched down, quickly rushed towards the shorebirds, set up a monocular telescope, and began to observe them carefully.

The gains this time was truly abundant. There were the Ruddy Turnstones with reddish-brown backs, white heads, and contrasting black breastbands that stood out. There were the Long-toed Stints with red crowns, fine black longitudinal stripes on their chests, and yellow legs. And there were the Sharp-tailed Sandpipers with red crowns, self-contained black V-shaped markings on their chests. All of them were new records for Hengqin Island.

横琴南部潟湖 Lagoon in the Southern Hengqin

# 2021 新年前夕横琴潟湖遇见黑脸琵鹭

黑脸琵鹭是全球濒危珍稀鸟种，对环境要求极为苛刻，是生态环境的指示性物种，其数量的增长与栖息地生态环境质量改善密切相关。在我国 2021 年 2 月 5 日发布的《国家重点保护野生动物名录》中，黑脸琵鹭从国家二级保护动物提升为国家一级保护动物。根据《黑脸琵鹭全球同步普查》结果，2023 年全球录得黑脸琵鹭的种群数量仅有 6603 只。

珠海市观鸟协会自 2013 年成立以来，连续 10 年，每年 1 月均组织义务调查员参与黑脸琵鹭全球同步普查活动，曾经在横琴岛的横琴码头至澳门大学沿岸红树林湿地、南湾红树林湿地、二井湾湿地、芒洲湿地、望眉角五个观测点附近均有黑脸琵鹭的观测记录。

横琴岛与澳门仅一水相隔，一衣带水，唇齿相依。珠海市观鸟协会与澳门生态环境保育协会一直保持良好的战略合作伙伴关系，历年来，均将横琴码头和澳门大学之间的水道，作为黑脸琵鹭的调查重点之一。

珠海市观鸟协会的郭会长与澳门生态环境保育协会的杨会长常常为黑脸琵鹭的归属地问题“争论不休”，鸟儿是会飞的，如何证明清晨在横琴岛觅食的黑脸琵鹭，夜晚不会飞到澳门栖息？

2021 年 2 月 7 日，阴历新年前夕，珠海市观鸟协会生态调查队进行野外调查时，在横琴潟湖发现了一大群白色鹭鸟当中，有几只低着头、长长的嘴巴扫来扫去的鹭鸟。“黑脸琵鹭！”郭会长立刻兴奋地大叫起来，通过单筒望远镜，观察到这群黑脸琵鹭共有八只，其中有一只戴有环志的黑脸琵鹭引起了关注，它的腿上佩戴着黄红色的鸟类环志，环志编号为 Y05。

通过检索查询，这是一只 2020 年 6 月 3 日在韩国 Bido 被环志的黑脸琵鹭。随着时间的推移，这只黑脸琵鹭逐渐向南迁徙，分别于 2020 年 10 月在福建省和 2020 年 11 月在澳门留下踪迹。接下来的观察记录则是 2021 年 2 月 7 日在珠海横琴。

这只 Y05 在横琴的现身，恰到好处地平息了郭会长与杨会长的“争执”。鸟儿是自由的，只要有适宜的生境、充裕的食物，它们都喜欢栖息及觅食。

其实，“抢鸟大战”只是一个乐子。近年来，两个协会精诚协作，在珠澳两地开展生态环保科研及宣教项目，向两地市民倡导绿色生态环保理念，深化两地环保宣教合作，共同呵护珠澳两地美好的生态家园。

# In 2021, Meet Black-faced Spoonbills in the lagoon of Hengqin on the eve of the New Year

Black-faced spoonbill is one of the endangered and rare bird species in the world. It is extremely demanding in terms of the environment and is an indicator species for the ecological environment. The growth of its population is closely related to the improvement of the ecological environment quality of its habitat. In China, the latest list of the national protected wild animals was officially released on February 5, 2021, and the Black-faced Spoonbill was upgraded from the Second Category of State Key Protected Wild Animals Species to the First Category species. According to the result of the *The International Black-faced Spoonbill Census Report 2023*, the population number of Black-faced spoonbill is 6603 globally.

Since its establishment in 2013, Zhuhai Bird Watching Society has organized voluntary investigators to participate in the International Black-faced Spoonbill Census in every January for ten years by 2023. The Black-faced spoonbills were recorded in five spots around Hengqin Island, which are mangrove wetlands along the coastline from Hengqin Pier to University of Macau, Nanwan Mangrove Wetland, Erjingwan Wetland, Mangzhou Wetland, and Wangmeijiao.

Hengqin Island and Macao is facing each other with only a waterway running between them. Hengqin Island and Macao are sharing a close relationship. Zhuhai Bird Watching Society and Macao Association of Ecological and Environmental Conservation have always maintained a good strategic partnership. Over the years, the waterway between Hengqin Pier and University of Macau has been regarded as one of the key sites for counting Black-faced Spoonbill.

President Guo of Zhuhai Bird Watching Society and President Yang of Macao Association of Ecological and Environmental Conservation often "dispute" over the ownership issue of Black-faced Spoonbill. Birds can fly, then how can we prove that Black-faced Spoonbill forages on Hengqin Island in the morning, and whether it will fly to inhabit Macao at night or not?

On February 7, 2021, the eve of the Lunar New Year, the ecological investigation team of Zhuhai Bird Watching Society conducted a field investigation and found a large flock of white herons in the southern lagoon of Hengqin, among which were several herons with their heads down and long beaks sweeping around. "Black–faced Spoonbill!" President Guo immediately shouted excitedly. Through the monocular, President Guo observed that there were 8 Black–faced Spoonbills in this group, and one of them with a bird banding on its leg drew her attention. It was a yellow–red bird banding numbered Y05.

A search revealed that this was a Black–faced Spoonbill ringed in Bido, South Korea on June 3, 2020. As time goes on, this Black–faced Spoonbill migrated southward, left tracks in Fujian Province in October 2020 and in Macao in November 2020. The following observation was recorded on February 7, 2021, in Hengqin, Zhuhai.

The appearance of Y05 in Hengqin just put an end to the "dispute" between President Guo and President Yang. Birds are free, and as long as there are suitable habitats and abundant food, they all like to live and forage there.

In fact, the "battle for the bird" was just for fun. In recent years, two associations have cooperated sincerely to carry out researches on ecological and environmental protection and education projects in Zhuhai and Macao, to promote the concept of green ecological and environmental protection to the citizens, to deepen the cooperation in environmental protection education, and to protect the beautiful natural homeland of Zhuhai and Macao jointly.

# 2021年广东新记录——靴隼雕翱翔脑背山

猛禽，泛指凶猛的掠食性或食腐性的鸟类，通常具有超强的飞行能力、粗壮有力的喙和脚爪，善于捕捉其他的小鸟及动物。它们处于食物链的顶层，是鸟类中的旗舰物种，其种类和个体数量远远少于其他类群，在自然生态食物链中对生态平衡调节起到重要作用，是关键性的、环境指示性物种。在我国重点保护野生动物名录中，所有的猛禽保护级别都是国家二级或以上，可见猛禽的保护至关重要。

每年的3—5月，是猛禽春季迁徙路过广东地区的时间。猛禽在迁徙途中需要在停歇地，即驿站停留一段时间，一般在2～13天，俗称“猛禽过境”。猛禽的栖息地比较特殊，多是远离人迹陡峭的山顶，攀登的危险系数大，较难到达此类猛禽栖息地进行观察；另外，猛禽飞行高度高、速度又快，也大大增加了观测的难度。

春季有部分猛禽沿海迁徙，会利用海岸边的适合生境。横琴岛处于东亚—澳大利西亚候鸟迁飞区，同时拥有海岸线和山林，容易形成上升气流，满足了猛禽利用上升气流盘旋进行迁徙的特性，因此，一些鹰形目、隼形目等类型的猛禽迁徙时会路过于此。

珠海市生态调查队特约技术指导廖晓东教授带领调查队考察横琴岛生境的时候，在脑背山上做过细致观察，认为该地海拔高度足够、视野宽广、气流合适，估计是猛禽过境的重要驿站，建议春季和秋季选择天气晴朗、吹北风或东北风3～4级的天气前来观测。

在横琴，可以观察到的猛禽除了常见的留鸟如黑鸢、凤头鹰、鹗等，还有机会观察到过境的猛禽如蛇雕、普通鵟、灰脸鵟鹰、红隼、燕隼等。

进入2021年春季，调查队把野外观测的重点转移到了猛禽身上。3月26日，调查队队员123在横琴岛进行鸟类调查时，发现周边小型雀形目鸟类突然叽叽喳喳吵闹不停，惊慌失措地飞起来四处逃窜，据他多年观测猛禽的经验判断，附近应该有掠食性猛禽。他立刻把目光投向天空，随后果然发现一只鹰形目猛禽在空中盘旋。123举起相机，将它拍摄下来，经过分析，这只猛禽是国家二级重点保护野生动物——靴隼雕。

123知道靴隼雕在我国新疆为夏候鸟，东北旅顺和河北为偶见冬候鸟。然而，他却从未在广东观测过它。经查阅《中国鸟类分类与分布名录（第三版）》、中国观鸟记录中心、“鸟语者”中国鸟类记录中心及相关历史文献，同时咨询中山大学生态学教授刘阳和广东省科学院动物研究所助理研究员张敏，均答复之前广东省未曾有过靴隼雕的记录。因此，可以确认本次记录的靴隼雕为广东省鸟类新记录，同时也是横琴岛的鸟类新记录。

# In 2021, A New Record in Guangdong — A Booted Eagle soars in Naobei Mountain

Raptor, generally referred to fierce predatory fed on carrions or other small animals, usually have super flying ability, thick and powerful beaks and claws, and are good at catching other small birds and animals. They are top predators of the food chain and are the flagship species of birds, which number of species and population of individuals are far less than those of other bird species. Raptors play an important role in balancing the ecosystem through the food chain and are key indicator species for the environment. In the list of national protected wild animals in China, all raptors are the first or second class of the national protected wild animals, which demonstrates the protection of raptors is very crucial.

From March to May, it is the time for raptors to migrate across Guangdong during their spring migration. During the migration, raptors need to stay at somewhere for food and rest for usually between two and thirteen days. Raptors like to stay at very rare habitat, mostly on the steep mountaintops far from humans, which makes it dangerous to reach and observe them; in addition, raptors soar fast and high in the sky, which also makes observation much more difficult.

In spring, some raptors migrate along the coast and will use suitable habitats along the coast. Hengqin Island is located within the East Asian – Australasian Flyway area, with mountain woods and a shoreline. The atmosphere around Hengqin can easily form rising column of air current to support raptors to soar and hover during migration. Therefore, some raptors, such as Accipitriformes and Falconiformes will pass Hengqin when migrating.

When Professor Liao Xiaodong, the advisor of the eco- investigation team of Zhuhai Bird Watching Society, led the team members to investigate the habitat of Hengqin Island, and he made a careful observation on Naobei Mountain. He believed it is a crucial habitat for migration raptors for its high altitude, vast space and suitable air current around. He suggested a sunny day with a northerly or northeasterly wind of 3 to 4 levels is good for spotting the birds of prey.

In Hengqin Island, not only commonly-seen resident raptors can be seen, such as Black Kite, Crested Goshawk, and Western Osprey, but also migratory raptors, like Crested Serpent Eagle, Eastern Buzzard, Grey-faced Buzzard, Common Kestrel and

Eurasian Hobby.

In the spring of 2021, the eco-investigation team focused at observing raptors. On March 26, the team member ABC noticed that a group of small passerine birds suddenly became noisy and flew around in panic during the eco-investigation in Hengqin. Based on his experience of raptor observation, ABC believed that there should be a predatory raptor nearby, and then an Accipitriformes raptor was found hovering in the sky. ABC immediately raised his camera and caught a photograph of it. He checked the photos carefully, and identified it is Booted Eagle, a raptor of the national second-class protected wild animal.

Booted Eagle is a summer migratory bird in Xinjiang, while an occasional winter migratory bird in Lushun and Hebei. However, it was the first time ABC spotted it in Guangdong. By looking up the information in "A Checklist on the Classification and Distribution of the Birds of China(Third Edition)", The China Bird Watching Record Center (www.birdreport.cn), "Bird Talker" and The China Bird Record Center (http://www.szbird.org.cn/birdtalker.net/index.asp) and relevant documents, and seeking advices from professor Liu Yang of Sun Yat-sen University and Zhang Min, an assistant researcher at the Institute of Zoology of the Guangdong Academy of Sciences, it turned out that there were no record of Booted Eagle in Guangdong ever before. Therefore, it could be confirmed that this record of Booted Eagle was a new bird record in Guangdong Province.

# 2021年横琴建设工地入夜对话林夜鹰

Chweep——Chweep——Chweep——一阵阵鸟类的鸣叫之声打破了入夜后逐渐寂静的横琴海洋乐园前的建设工地，这声音立刻引起了正在进行夜间野外调查的珠海市观鸟协会生态调查员小鱼儿的注意，借着远处微弱的路灯光亮，隐隐见到一只飞行动物在工地上低空盘旋。

小鱼儿叫上同行的阿棠，深一脚浅一脚地循着声音的来源、模糊的影子往工地的深处追去。快走到飞行动物十几米的位置，它忽然一个转身，蹿往更远处隆起的土堆后面，隐藏不见了，鸣叫声也随之消失。

他俩坐在工地的土堆上，准备休息一下喝口水，一个黑影突然再次 Chweep——Chweep——Chweep——大叫着从远处冲着他们飞来，阿棠赶紧扔下水杯举起手电筒，小鱼儿则拿起相机一顿咔嚓咔嚓猛拍，可惜周围实在太暗，手电筒亮度不够，加上夜拍经验不足，只拍到了模糊的照片，依稀能够辨认出这是林夜鹰。

“黑夜给了我黑色的眼睛，我却用它寻找光明。”夜间，在包括人类在内的大部分生物进入梦乡酣睡之时，有一些夜行性动物却异常活跃。林夜鹰就是活动在夜里的积极分子，它们在夜里以捕捉空中飞行的各种昆虫为食，一双大眼在黑暗中闪闪发亮，然而它的羽色和树皮非常相似，具有很好的保护色，人们极难发现。

为了拍好林夜鹰，2021 年 5 月 26 日— 6 月 4 日，连续 10 个夜晚，调查队队员们不辞辛劳奔赴那片工地，守候与林夜鹰的相会，一守就是数个小时，常常被蚊虫叮咬得苦不堪言。有一天晚上雨后泥泞，小鱼儿不慎滑倒，起来时变成了“黄泥鱼”。

郭会长有个绝活儿，就是模仿鸟儿的叫声。于是，原本寂静的建设工地夜晚，便时不时地响起清脆的鸟鸣之声。正是这样的刻苦与执着，打动了骄傲的林夜鹰，它时而悄无声息地跟大伙儿躲猫猫，时而又好奇地循着郭会长的声音，大声叫唤着在大伙儿附近飞翔，仿佛在跟她亲切地对话。

每当林夜鹰飞近之时，郭会长、林中少帅、柯竹、小鱼儿赶紧举着相机追拍。夜太黑，刚开始的拍摄效果很不理想，不是追踪不上，就是成像模糊不清。大家不断总结经验，调整相机参数，一次又一次地尝试，终于在团队的紧密合作之下，拍摄到清晰而又珍贵的林夜鹰夜行的照片。

# In 2021, Dialogue with the Savanna Nightjar in the Construction Site of Hengqin at Night

Chweep—Chweep—Chweep—The sound of bird calls breaks the gradual quiet of the construction site in front of the Hengqin Ocean Park, as night falls. This sound immediately catches the attention of Little Fish, the eco-investigator of the Zhuhai Bird Watching Society, who is conducting nighttime field research. With the faint glow from a distant streetlight, they can vaguely make out a flying creature, hovering low over the construction site.

Little Fish called his colleague Tang, and together they carefully followed the source of the sound and the blurry shadow by stepping deeper into the construction site. They moved cautiously step by step, sometimes sinking into the ground, as they pursued the source of the sound. As they approached a distance of about a dozen meters from the flying creature, it suddenly made a swift turn and dashed towards a raised mound of soil further away, disappearing from sight. The calls also faded away, silence taking over once again.

They sat on the mound of soil at the construction site, preparing to take a break and have a sip of water. Suddenly, a black shadow once again gave the Chweep—Chweep—Chweep—call from a distance and flew towards them. Tang quickly dropped the water cup and raised the flashlight, while Little Fish grabbed the camera and started snapping photos with a rapid clicking sound. Unfortunately, the surroundings were too dark, and the flashlight wasn't bright enough. Coupled with their limited experience in night photography, they only managed to capture blurry photos. However, they could still discern the distinct features of the bird, indicating that it was the Savanna Nightjar.

"The dark night has given me black eyes, but I use them to seek the light." During the night, when most creatures, including humans, enter a deep sleep and dreamland, some nocturnal animals become exceptionally active. The Savanna Nightjar is one such enthusiastic creature that thrives in the night. It hunts various insects that fly through the air, relying on its keen vision. Its large eyes glisten in the darkness. However, its tree-bark-like-dark-brown feather color acts excellent camouflage in the woods that makes it extremely difficult to spot.

In order to capture the perfect shot of the Savanna Nightjar, from May 26th to June 4th, 2021, the members of the research team tirelessly ventured to the construction site for ten consecutive nights. They patiently awaited the encounter with the Savanna Nightjar, often spending several hours on watch. They endured the torment of relentless mosquito bites. One night, after rainfall, the ground became muddy. Unfortunately, Little Fish slipped down and covered in yellow mud, making him like "Muddy Fish".

President Guo has a special talent: imitating bird calls. As a result, the once silent construction site would occasionally be filled with the clear and melodious sounds of bird songs during the night. It was through this dedication and perseverance that the proud Savanna Nightjar was moved. It silently plays hide and seek with everyone at times, or curiously follows President Guo's calls at times and loudly calling and flying around, as if engaging in a friendly conversation with President Guo.

Whenever the Savanna Nightjar approached, President Guo, "The Marshal", Kezhu, and Little Fish would quickly raise their cameras to capture its figure. The darkness of the night posed a challenge, and the initial shooting results were not ideal. They either struggled to keep up with the Savanna Nightjar's movements or ended up with blurry images. However, through continuous learning and adjustment of camera settings, they persevered and made numerous attempts. Finally, with the close cooperation of the team, they successfully captured clear and precious images showing the night-time activity of the Savanna Nightjar.

# 2021 中秋前夕望洋角白腹海雕搏击长空

白腹海雕，国家一级保护动物。作为珠海市海岛生境的代表猛禽，其高冷的外表霸气外泄，堪称野生猛禽鸟类中的“战斗机”。

珠海市拥有 146 个岛屿，号称“百岛之市”，有着丰富的海岛资源。2013 年始，珠海市观鸟协会对有人居住的各大岛屿开展了海岛野生鸟类资源调查。经过九年多的持续追踪调查，在珠海市万山群岛的外伶仃岛、担杆岛、东澳岛、大万山岛、桂山岛均有白腹海雕的记录，其中外伶仃岛和担杆岛还有它们的繁殖记录。而距离珠海市区较近的荷包岛、淇澳岛，甚至在市区的高栏港、香洲区沿海岸线的大镜山水库也有白腹海雕曾经出现的记录，唯独在横琴岛没有见过它的踪迹，这在郭会长心中不得不说是一个遗憾。

2021 年 9 月 19 日，生态调查队对横琴岛进行野外巡护调查。尽管已接近中秋节，当天的“秋老虎”对珠海进行反扑，气温高达 35℃，却是晴空万里，天空中飘着朵朵白云，去横琴岛调查的路上，景色美不胜收。

林中少帅凭借高超的车技，穿过沟沟坎坎的施工工地以及算不上路的泥巴土道，一路颠簸抵达望洋角海边，刚刚下车准备舒展一下筋骨，突然一声大喝：“猛禽！”然后，一个反扑回到车里，抄出单反“大炮”对着天空猛拍。回过神来的郭会长、柯竹也毫不示弱，立刻抱着相机跳下车。只见一只猛禽，在头顶 20 多米的空中盘旋，一边翱翔一边用锐利的目光紧紧盯住下方突然出现的人们，好像它也在奇怪，如此难走的道路且人迹罕至的地方，怎么会有人类这样的庞然大物出现？

低空盘旋了数圈之后，白腹海雕盘旋上飞，在高空又翱翔了数圈，才依依不舍地朝着南边远处的海岛方向越飞越远。

依依不舍地告别白腹海雕，三个人各自欣喜若狂，林中少帅感叹道：“在外伶仃岛上拍摄白腹海雕那么多年，一直都是在近百米之外的山崖之上，从来没有如此近距离地见过它！刚下车的时候，它停在不远的树上，距离可能也就是十多米，一个庞然大物突然惊飞起来掠过头顶，颇为壮观啊！”柯竹一边回看照片，一边激动地说：“天哪！太高清了，竟然连它的舌头都看得清清楚楚！”郭会长却指着远处的海岛，感慨万千：“你们看到了？！这么多年，我一直在说，横琴岛与万山群岛在海域上连成一片，有山、有峭壁、有海岸线，完全符合白腹海雕生存的环境，它们肯定会造访这里！看，被我说中了吧！！！”

此次发现的白腹海雕，更新了横琴岛鸟类记录，也填补了横琴岛没有白腹海雕记录的空白！

# In 2021, A White-bellied Sea Eagle hovering at Wangyangjiao on the eve of Mid-Autumn Festival

White-bellied Sea Eagle is the first class of the national key protected wildlife. As the representative raptor of the island habitat of Zhuhai, and due to its fierce appearance, it is considered as a "fighter" among wild raptors.

Zhuhai owns 146 islands, known as the "City with Hundreds of Islands", and enjoys rich island resources. Since 2013, Zhuhai Bird Watching Society has carried out investigations of wild bird resources on the inhabited islands. After more than nine years of continuous tracking and investigation, we have records of White-bellied Sea Eagle on Wailingding Island, Dan'gan Island, Dong'ao Island, Dawanshan Island and Guishan Island in the Wanshan Islands group of Zhuhai, and breeding records on Wailingding Island and Dan'gan Island. Moreover, there were also records of White-bellied Sea Eagle on Hebao Island and Qi'ao Island, where are close to the downtown, and even along the coastline area of Xiangzhou District, like the Gaolan Port and Dajingshan Reservoir. But no records were recorded in Hengqin, which was a pity in President Guo's mind.

On September 19, 2021, the eco-investigation team was conducting a field patrol on Hengqin Island. Although the Mid Autumn Festival was approaching, the weather temperature is as high as 35 Celsius degree in Zhuhai. What a sunny day! White cloud was floating in the sky, and the scenery was very beautiful on the way to Hengqin Island for eco-investigation.

With his superb driving skill, the Marshal, one of our team members, drove through the gully and muddy path in the construction site, which was not regarded as a road. He bumped all the way to the seaside at Wangyangjiao. When he just got off and stretch himself, he suddenly shouted, "A raptor!" Then he turned back and dived into the vehicle, took out his camera and caught photographs against the sky. President Guo and the team member Bamboo had finally realized what had happened, and immediately jumped out of the vehicle with their cameras as well. A raptor was hovering overhead, about 20 meters high. While hovering, it stared with sharp eyes at the people who suddenly appeared below, and it was also wondering why human beings could be in such a wild place with difficult path.

After soaring for several circles at low altitude, the White-bellied Sea Eagle hovered up again at high altitude, and then flew away gradually further and further towards the island far to the south.

Reluctantly bid farewell to the White-bellied Sea Eagle, they were all ecstatic. The Marshal sighed:"I have been photographing White-bellied Sea Eagle on Wailingding Island for so many years, they were all on a rocky cliff nearly 100 meters away, and I have never seen it so close! When I got off my vehicle, the eagle stopped on a tree only about 10 meters away. A giant creature suddenly flew over my head, quite a spectacular!" Bamboo looked back at the photographs and said excitedly:"Oh! It's so high-definition that even its tongue can be seen clearly!" However, President Guo pointed to the island in the distance and said emotionally: "See! For so many years, I have been saying that Hengqin Island and the Wanshan Islands are connected in the sea area, with mountains, cliffs and coastline, which completely conforms to the living environment of White-bellied Sea Eagle. It would definitely visit here! You see, I was right!!!"

This discovery of White-bellied Sea Eagle renewed the list of bird record on Hengqin Island, and filled the gap of no White-bellied Sea Eagle recorded on Hengqin Island.

# 2022 年神秘的赭红尾鸲造访小横琴山

2022 年 5 月 17 日，珠海市观鸟协会生态调查队在小横琴山进行日常调查，吃完自带的简单午餐后，一行人在林荫树下小憩，林中少帅忽然听见杜鹃科鸟类的叫声，看到周围植被茂盛，决定放弃休息在周边走走继续鸟类观察。

当林中少帅走到不远处的苗圃时，一只特别的鸟儿闯入他的视野，他眼疾手快地举起相机开始拍摄，鸟儿似乎发现有人在偷拍它，立刻飞进树丛躲了起来。“快来，有好鸟啊！”他赶紧打电话把消息分享给了同行的队员。

郭会长和凌紫闻讯立刻赶来，这时的鸟儿似乎也适应了人们的出现，从树丛中飞出来，时而跳进苗圃寻找食物，时而跃上枝头左顾右盼，还俏皮地歪着头与凌紫对望，可把凌紫逗乐了，忍不住疑惑地问：“你是谁？从哪里来？又准备往哪里去？”

看似简单的问题，却引发了一番激烈的争论。为了确认这只鸟儿的身份，秉着科学严谨的态度，郭会长、林中少帅、凌紫，叫上小鱼儿一起，次日再次来到该地进行观察拍摄，鸟儿似乎也把这里当成了歇脚的宝地，欢快地在苗圃里蹦蹦跶跶地迎接调查队员们，通过多角度的观测、对比鸟类图鉴、咨询鸟类专家，最终确认该鸟为赭红尾鸲。

赭红尾鸲的身份确认了，然而资料显示，它在广东省只有过文字记录，具体影像资料及确切位置存疑。本次发现和拍摄的赭红尾鸲高清影像，为广东省赭红尾鸲的记录贡献了确切的影像证据，这也是珠海市首次发现赭红尾鸲的新记录。

那么问题还没完全解决，它从哪里来？又准备往哪里去呢？可以肯定的是，赭红尾鸲是一种有迁徙行为的候鸟。迁徙是鸟类遵循大自然环境的一种生存本能反应，它们每年春秋两季沿着固定的路线往返于繁殖地和避寒地之间，不畏艰辛，无惧风险，只为了生存的本能而飞翔。为了早日抵达目的地，候鸟们需要日夜兼程，资料显示，有一只候鸟，用 10 天时间飞越 13000 多公里，是候鸟的连续飞行记录。

这只意外造访小横琴山的赭红尾鸲，只是短暂停留了两天，此后数日再去寻找，它已不知所终。好吧，姑且放下这一不小心就触及的苏格拉底“哲学三连问”，对于候鸟迁徙之谜的探索，研究候鸟的迁徙时间、路线、繁殖地、越冬地环境的关系等生态规律，对于保护珍稀濒危鸟种、维护生态平衡可以提供科学的依据，将会带来良好的社会效益和生态效益，为此，珠海市观鸟协会一直在努力。

# In 2022, The Mysterious Black Redstart Visits Lesser Hengqin Mountain

On May 17, 2022, the Zhuhai Bird Watching Society's eco-survey team conducted their routine survey on Lesser Hengqin Mountain. After finishing their simple packed lunch, the team took a short break under the shade of the trees. Suddenly, one of the team members, "The Marshal", heard bird calls of the Cuculide family. Seeing the lush vegetation around, they decided to forgo their rest and explore the nearby area for birds.

As "The Marshal" approached a nearby plant nursery, a special bird suddenly appeared in his field of vision. Reacting quickly, he raised his camera and began capturing photos. The bird seemed to notice that it was being photographed and immediately flew into the thicket to hide. "Come quickly! There's an amazing bird!" "The Marshal" hurriedly called his fellow team members to share the exciting news.

President Guo and Lingzi immediately rushed to the spot upon receiving the news. the bird seemed to have become accustomed to the presence of humans by this time. It flew out of the thicket, hopping into the plant nursery to search for food. It then leaped onto tree branches, looking around curiously. Playfully, it tilted its head and made eye contact with Lingzi, which delighted her. Unable to contain her curiosity, Lingzi asked, "Who are you? Where do you come from? And where are you planning to go?"

What seemed like a simple question sparked a heated debate. In order to confirm the species of the bird, and with a commitment to scientific rigor, President Guo, "The Marshal", Lingzi, and Little Fish returned to the site the next day for further observation and photography. The bird seemed to consider this place as a resting haven and cheerfully greeted the survey team in the plant nursery. Through multiple angles of observation, comparison with several bird field guide books, and consultation with bird experts, they ultimately confirmed that the bird was indeed a Black Redstart.

The identification of the Black Redstart has been confirmed. However, according to available information, it has only been recorded in written form in Guangdong Province, but with certain photographic evidence and exact locations. The high-resolution images of the Black Redstart discovered and captured during this observation contributed as concrete visual evidence for the record of this species in Guangdong Province. This finding fulfilled the perilous gap, and the finding also marked a new record of the Black Redstart in Zhuhai City.

The question of where the Black Redstart comes from and where it is headed is still not fully resolved. However, it can be confirmed that the Black Redstart is a migratory bird. Migration is a survival instinctive response followed by birds in accordance with the natural environment. They travel along fixed routes between their breeding grounds and wintering grounds during the spring and autumn seasons. They endure hardships and face risks, all driven by their instinct for survival. To reach their destination as quickly as possible, migratory birds need to travel day and night. According to available information, there is a migratory bird that flew over 13,000 kilometers in just 10 days, setting a record for continuous bird flight.

The Black Redstart that unexpectedly visited Little Hengqin Mountain only stayed for two days. It could not be found in the follow days. Well, let's set the Socratic "three-question philosophy" aside for now. Exploring the mysteries of bird migration and studying the ecological patterns related to the timing, routes, breeding grounds, and wintering habitats of migratory birds can provide a scientific basis for protecting rare and endangered bird species and maintaining ecological balance. This endeavor will bring about positive social and ecological benefits. The Zhuhai Bird Watching Society has been striving towards the goal.

# 2023 年石栏洲风雨中舞蹈的大滨鹬

2023 年 3 月的一天下午，气象预警信息提示海面有 8 级风力及黄色暴雨警报，珠海市观鸟协会生态调查队驱车来到了横琴石栏洲，远远望去，青黑色的乌云阴沉沉的仿佛要坠落海面，海风凌厉地叫嚣着，往昔平静的“海豚湾”沙滩上翻涌起一层层的海浪，呈现出一幅暴风雨即将来临的水墨画卷。

晓乔疑惑地望着整装待发的郭会长，小声地嘟囔：“这样的坏天气，还要去海边做巡护吗？”郭会长斩钉截铁地说：“去，我们快去快回，越是这样的天气，越有可能出现‘妖怪’呢！”

郭会长、林中少帅、晓乔带着装备，小心翼翼地穿过海边的乱石堆，来到了“海豚湾”的沙滩上，天气不好能见度低，郭会长架起单筒望远镜，沿着海岸线逐一扫描，突然，她低声喝了一句：“在我前方大约 60° 的方向，有一小群鸻鹬类的鸟，走路摇摇晃晃的像跳舞一样。”

林中少帅和晓乔抱着相机、猫着腰，慢慢地朝着郭会长指的那群小黑点靠近，远远的距离林中少帅按动了快门，放大了照片仔细辨认，那一群鸟儿，竟然是世界自然保护联盟 IUCN 所列濒危物种、国家二级保护动物——大滨鹬！晓乔在旁边心急火燎：“我的镜头太短了，拍不清楚啊！”林中少帅说：“别着急，为了避免鸟儿惊飞，咱们首先要远远地拍照记录下来，再逐渐靠前，争取拍到更加清晰的影像！”

如此阴郁、能见度差的天气，大滨鹬们似乎也放松了警惕，自顾自地在沙滩上踏着舞步觅食、玩耍，还有几只扑腾到浪里来一段水上芭蕾，再逐渐靠近，晓乔也拍摄到了比较满意的照片，认真数了数，竟然有 23 只。

天边的乌云越压越低，开始飘起了零星的雨点，“快撤！”林中少帅果断地叫上郭会长和晓乔往车的方向奔去，刚刚上车，瓢泼大雨倾盆而下，周围的一切瞬间被雨帘包裹着，唯有车成为躲避风雨的港湾。

晓乔缓了缓神，叹道：“咱们可以躲在车上遮风避雨，可是大滨鹬们要去哪儿藏起来呢？郭会长，我特别想采访一下你此时此刻的心情如何呢？”

郭会长微微一笑，淡定地说：“这样的风雨算不了什么？！鸟儿们的生存智慧对于人类来说，存在着很多的谜团。正是一个又一个为什么，支持着我们鸟会十年如一日不懈地追寻和探索鸟儿们的世界，经历大风大雨是常有的事情，我们早就习惯了，这样的天气，遇到特别的鸟是一种缘分，遇不上也没啥遗憾的。”

调查队返回后查询资料，这次风雨中偶遇的大滨鹬，是横琴岛的新记录。

# In 2023, The Great Knot dancing in the storm on Shilanzhou Island

In the afternoon of one day in March 2023, the weather alert warned of a level 8 wind and a yellow rainstorm warning over the sea. The eco–investigation team of the Zhuhai Bird Watching Society drove to Shilanzhou Island in Hengqin. From a distance, dark and gloomy clouds seemed to loom over the sea, and the fierce sea wind howled. The once calm "Dolphin Bay" beach was now filled with layers of turbulent waves, creating an ink painting that portrayed an imminent storm.

"Do we have to go on patrol at the seaside in such bad weather?" Xiaoqiao asked President Guo, who was packing up and getting ready. President Guo replied firmly, "Yes, we'll go and come back quickly. The more the weather is like this, the more likely it is for 'monsters' to appear!"

President Guo, "The Marshal", and Xiaoqiao cautiously made their way through the scattered rocks by the seaside and arrived at the sandy beach of "Dolphin Bay." The weather was bad, with low visibility. President Guo set up her monocular telescope and scanned along the coastline. Suddenly, she whispered, "About 60 degrees ahead of me, there's a small group of sandpipers. They're walking unsteadily, almost like dancing."

"The Marshal" and Xiaoqiao, holding their cameras and crouching low, slowly approached the group of small black dots pointed out by President Guo. From a distance, "The Marshal" pressed the shutter button and enlarged the photos to examine them closely. To their astonishment, the birds in the group turned out to be the endangered species listed by the IUCN and in the list of National Secondary Protection of Wild Animals—the Great Knot! Xiaoqiao cried out anxiously, "My lens is too short, I can't capture a clear image!" "The Marshal" reassured her, saying, "Don't worry. To avoid startling the birds, we should first capture some distant photos and then gradually get closer, aiming to obtain clearer images!"

Despite the gloomy weather and poor visibility, the Great Knots seemed to have relaxed their alertness. They were busy searching for food and playing on the sandy beach, with a few of them splashing into the water for a ballet. As they approached closer, Xiaoqiao also captured some satisfied photos. She carefully counted them and was amazed to find a group of 23 individuals.

As the dark clouds on the horizon grew lower, sporadic raindrops began to fall.

"Let's retreat!" "The Marshal" decisively called out to President Guo and Xiaoqiao, and they rushed towards the direction of their car. Just as they got into the car, the rain is pouring down like a torrent. Everything around is instantly wrapped in the curtain of rain, with only the car becoming a shelter from the storm.

Xiaoqiao took a moment to gather her thoughts and sighed, "While we can seek shelter in the car to shield ourselves from the wind and rain, I wonder where the Great Knots find refuge. President Guo, I'm particularly curious to hear about your thoughts and feels at this very moment. Can I interview you?"

President Guo smiled and replied calmly, "This storm is nothing. The survival wisdom of birds holds many mysteries for humans. It is these endless 'whys' that drive our members to tirelessly pursue and explore the world of birds year after year. Enduring strong winds and heavy rain is a common practice for us, and we have long become accustomed to it. In weather like this, encountering a special bird is a matter of fate, but even if we don't come across one, there is no regret".

After returning, the research team checked the records and found that the encounter with the Great Knot during the storm was a new record for Hengqin Island.

# 附录
# Appendix

# 鸟类基础知识

## · 鸟的形态

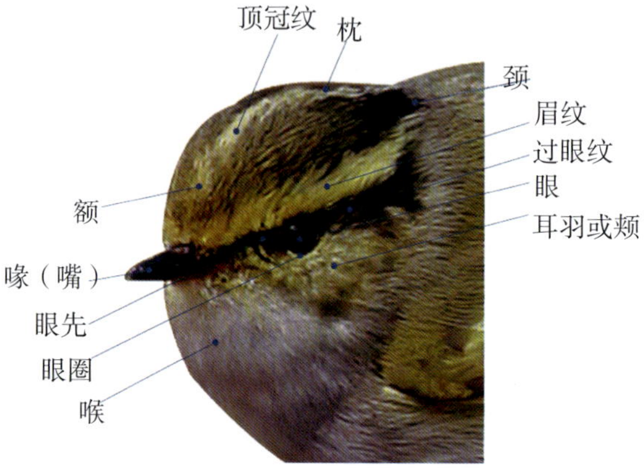


鸟类头部各部位名称

鸟类身体各部位名称

## · 鸟的居留类型

由于气候、日照时间、温度、食物等的变化，以及鸟类内在生理因素的影响，一些鸟类会进行季节性迁徙。根据是否进行迁徙，可以将鸟类分为不同的居留类型。

留鸟：终年栖息在同一地区并在本地繁殖，不进行远距离迁徙的鸟类。比如，麻雀、鹊鸲等。

候鸟：春秋两季沿着比较固定的路线在繁殖地和越冬地之间进行迁徙。比如，赤颈鸭、灰鸻、反嘴鹬等。

在不同的地域，根据候鸟出现的季节，又细分为夏候鸟、冬候鸟、过境鸟。

这样的划分是有地域性的。春季到达某地，夏季在繁殖，秋季离开，该鸟就是此地的夏候鸟。秋季迁徙，冬季到某地越冬，春季飞回繁殖地，该鸟则是此地的冬候鸟。迁徙途经某地，只是短暂停留，补充食物后继续迁徙，不在该地繁殖或越冬，此鸟就是过境鸟，或者旅鸟。

迷鸟：指那些在迁徙过程中由于狂风等某种特殊环境、气候因素而偏离正常的迁徙路径，而偶尔到异地的鸟类。

## · 鸟类保护等级

国一：国家一级保护野生动物。

国二：国家二级保护野生动物。

三有：国家保护的有重要生态、科学、社会价值的陆生野生动物。

世界自然保护联盟（IUCN）红色名录：濒危（EN）、易危（VU）、近危（NT）、无危（LC）、数据缺乏（DD）。

## · 鸟类观察及鸟类摄影守则

1. 观鸟须知

服装避免太鲜艳，宜选择与自然环境协调的颜色。

保持安静，轻声交谈，动作轻缓，与鸟类保持适当距离。

准备好望远镜，选择适合的图鉴辨认鸟类，认真观察，做好笔记。

注意安全，防备危险的动植物等。

避免接近危险区域或误入禁止区域。

发现野鸟及时告诉同伴，对初学者给予帮助。

避免干扰他人，让大家都可以享受观鸟乐趣。

2. 拍鸟须知

应用自然光，少用闪光灯，以免惊吓雀鸟。

不过分追赶野鸟，拍摄时，应保持一定距离。

不可为了便于拍摄，随意攀折花木，破坏野鸟栖息地以及附近植被生态，或干扰雀鸟，如使用诱饵、放鸟鸣录音带或丢掷石头惊吓驱赶等行为。

慎重拍摄鸟巢，禁止拍摄繁育中的珍稀鸟类。

珍稀鸟种摄影照片慎重发表，切勿透露拍摄地点，保护其栖息地信息。

3．绿色出行

如发现非法捕鸟或伤害野生鸟类的行为，在保证自身安全的前提下，拍照记录，并尽快报警举报。

未经许可，不采集野生动植物标本或活体。

不要破坏当地的设施或农作物。

尽量减少携带瓶装水、饮料、袋装食物等，避免产生过多的一次性垃圾。

保护环境，不乱扔垃圾。

## · 鸟类保护常识

1. 请不要随意喂鸟

很多市民看到野生鸟类，都喜欢拿着人类的食物给它们喂食，或是为了乐趣，或是出于好心帮它们解决食物问题，但其实随意投喂鸟类会导致许多问题。

人类的食物对鸟类来讲很可能是不健康的，鸟类有适合它们的饮食，如果喂错食物可能会导致营养不良或死亡。喂食会改变动物的觅食习性，使鸟类产生依赖性，一旦离开人类就无法生存。近距离喂食和接触鸟类也可能会增加人与动物之间疫病相互传播的风险。所以为了保护人类及鸟类的生命健康，请您不要随意喂食野生动物，也请尊重它们的生活方式。

### 2. 路边的小鸟请不要随意捡

如果在路边遇到刚出生的鸟宝宝，不要随意捡回家，先观察鸟爸鸟妈是否在身边，如在身边，就不用管，它们最懂得如何照料自己的宝宝。

如未发现成鸟在周边，且小鸟受伤或无法飞行，请及时联系野生动物保护机构。如果小鸟状态良好，请在保证自身安全的情况下，戴好防护手套找到附近的鸟巢，把小鸟放回去就好。

### 3. 发现非法张网捕鸟或打鸟怎么办

市民在登山、郊游或日常生活中，如发现有人张网捕鸟或打鸟，请在确保自身安全的情况下，及时拨打 110 报警电话举报。

# Knowledge of Avian

## · Morphology of a Bird

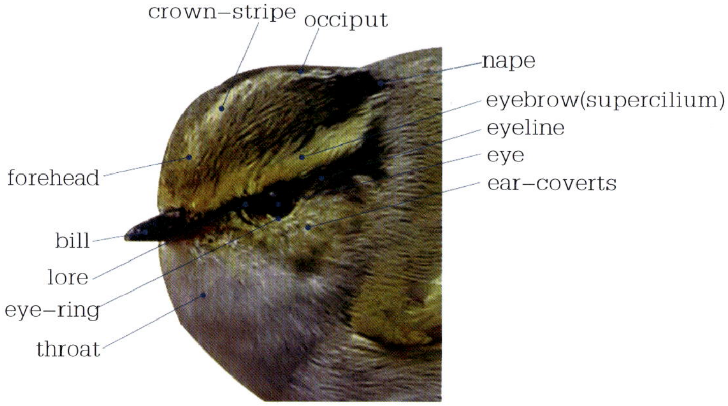


Parts of a Bird Head

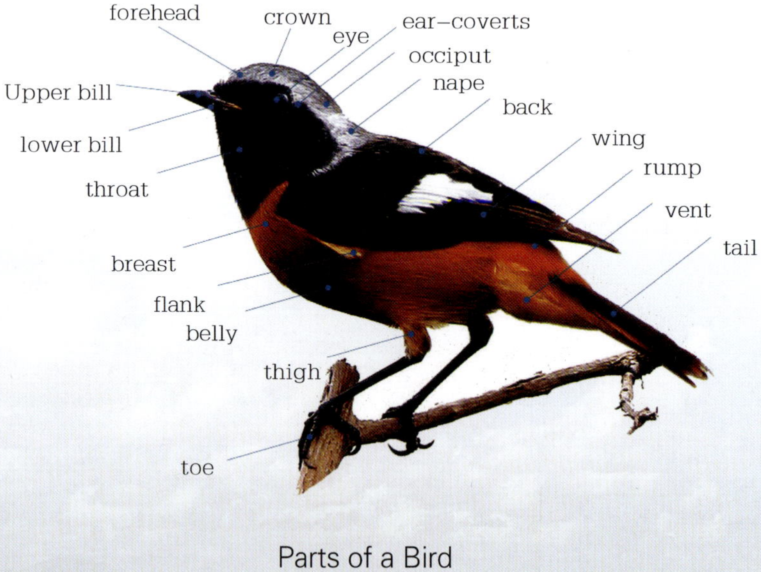


Parts of a Bird

## · Residency Category of Birds

Birds will take seasonal migration due to changes of weather temperature, day length, food availability and the impacts of their physiological metabolism. Birds can be identified as migratory and resident birds.

RESIDENT BIRD: Resident birds are birds that stay the whole year within their breeding range. They would not take long distance migration regardless of food availability or even because of the harsh weather conditions. In Zhuhai, such as Eurasian Tree Sparrow and Oriental Magpie Robin are resident birds.

MIGRATORY BIRD: Due to seasonal changes, migratory birds will fly back and forth along regular routes between breeding areas and wintering grounds. Such as Eurasian Wigeon, Grey Plover and Pie Avocet are winter migratory birds in Zhuhai.

Based on the seasons when migratory birds appear, they are further classified into summer visitors, winter visitors, and passing visitors.

There are many different migration patterns. If a bird arrives in a certain area in spring, breeds during summer, and departs in autumn, it is considered a summer bird in that area. If a bird arrives in autumn, spends the winter in a specific area, and flies back to its breeding grounds in spring, it is known as a winter bird in that location. Birds that pass through a certain area, only stopping to replenish themselves with food before continuing their migration but without breeding or wintering there, are referred to as passing visitor.

VAGRANT BIRD: It refers to those birds that deviate from their normal migration routes due to strong winds or other environmental or climatic factors during migration, and occasionally end up in unusual migratory locations.

## · Protection Categories of Birds

Ⅰ: National First-class Protected Wild Animals

Ⅱ: National Second-class Protected Wild Animals

3V: ("3 Values" for short name) A national list of terrestrial wildlife of important ecological, scientific, and social values.

International Union for Conservation of Nature (IUCN) Red list categories:

EN: Endangered

VU: Vulnerable

NT: Near Threatened

LC: Least Concern

DD: Data Deficient

## · Code of Conduct for Birdwatching and Bird Photography

### 1. Guidelines for Birdwatching

Clothing: Avoid bright and flashy colors. Choose colors that fit in the natural environment.

Maintain silence: Speak softly and move slowly. Keep an appropriate distance from the birds.

Preparation: Binoculars, Bird guides. Observe carefully and take notes.

Be aware of safety: Watch out for potentially dangerous animals and plants.

Avoid restricted areas and no enter to dangerous or prohibited areas.

Share information: Notify your fellow birdwatchers if you spot any interesting birds. Offer assistance to beginners.

Respect others: Avoid disturbing others and ensure everyone can enjoy birdwatching.

### 2. Guidelines for Bird Photography

Use natural light and avoid excessive use of flash to prevent startling the birds.

Do not excessively chase after wild birds. Maintain a certain distance when taking photographs.

Do not damage bird habitat environment for the convenience of photographing, such as breaking branches or disturbing birds by using bait, playing birdcall recordings, or throwing stones.

Exercise caution when photographing bird nests and refrain from photographing rare bird species during breeding.

Exercise discretion when publishing photographs of rare bird species, and avoid disclosing specific locations in order to protect the habitat information.

### 3. Green Travel

If you witness any illegal bird trapping or damage to wild birds, take photos as evidence, and report to the authorities as soon as possible while ensure your own safety first.

Do not collect specimens or live wild animals and plants without permission.

Do not damage local facilities or crops.

Minimize the use of bottled water, beverages, and packaged food to reduce the generation of excessive single-use waste.

Protect the environment and refrain from littering.

## · Bird Conservation Knowledge

### 1. Please do not feed birds indiscriminately

Many people enjoy feeding wild birds when they encounter them. Whether it's for fun or to help them find food, random bird feeding can lead to several problems.

Human food is often unhealthy for birds, as they have specific dietary needs. Feeding them the wrong food can cause malnutrition or even death. Feeding can also alter their foraging behaviors, making them dependent on human-provided food and unable to survive without it. Additionally, close contact and feeding birds at a short distance can increase the risk of disease transmission between humans and birds. To protect the health and well-being of both humans and birds, it is important to refrain from feeding wild animals and respect their natural way of life.

### 2. Please do not pick up birds by the roadside indiscriminately

If you encounter newly hatched bird babies by the roadside, do not pick them up and take them home without observation. First, observe if the bird parents are around. If they are, there is no need to intervene as they are the best caretakers for their own babies.

If you do not see adult birds nearby and the baby bird is injured or unable to fly, please contact a wildlife conservation organization promptly. If the baby bird is in good condition, with proper protective gloves and ensuring your own safety, locate the nearby bird nest and place the baby bird back inside.

### 3. What to do if you encounter illegal bird trapping or bird shooting

If you come across someone engaging in illegal bird trapping or bird shooting while hiking, on an outing, or in your daily life, please ensure your own safety and promptly call the emergency hotline 110 to report the incident.

# 横琴鸟类名录索引

# Hengqin Bird List Index

## · 中文名称索引

## · Index of English Names

# 主要参考文献


[1] 邹发生等 . 广东陆生脊椎动物分布名录 [M]. 广东科技出版社，2016.

[2] 马敬能，何芬奇等 . 中国鸟类野外手册 [M]. 湖南科学技术出版社，2000.

[3] 刘阳，陈水华 . 中国鸟类观察手册 [M]. 湖南科学技术出版社， 2021.

[4] 国家林业和草原局 . 国家重点保护野生动物名录 . 2021.

[5] 广东省林业局 . 广东省重点保护陆生野生动物名录 . 2021.

[6] 华南濒危动物研究所. 珠海市横琴岛动物调查报告 . 2011.

[7] 横琴本岛第七次全国人口普查公报（第一号）-- 常住人口情况

(http://www.hengqin.gov.cn/stats/tjsj/rk/content/post_3495703.html )

# References


[1] Zou Fasheng et al. Guangdong Terrestrial Vertebrates Distribution List. Guangdong Science & Technology Press. 2016.

[2] John MacKinnon, He Fenqi et al. A Field Guide to the Birds of China. Hunan Science & Technology Press. 2000.

[3] Liu Yang, Chen Shuihua. The CNG Field Guide to the Birds of China. Hunan Science & Technology Press. 2021.

[4] National Forestry and Grassland Administration. The National Protected Wild Animals List. 2021.

[5] Forestry Administration of Guangdong Province. The Guangdong Provincial Key Protected Terrestrial Wild Animals List. 2021.

[6] Research Institute of Southern China Endangered Animals. Investigation Report of Terrestrial Wild Animals in Hengqin Zhuhai. 2011.

[7] The No.7 National Population Census Report in Hengqin.

(http://www.hengqin.gov.cn/stats/tjsj/rk/content/post_3495703.html)

# 珠海市观鸟协会简介

珠海市观鸟协会成立于2013年11月，是珠海市民政局登记注册的公益类社会组织，同时是广东省湿地协会团体会员单位、珠海市科学技术协会团体会员单位、珠海市科普联盟成员单位。2019 年 4 月，荣获珠海社会组织评估等级 5A 级称号。

协会秉承“人与自然和谐”的愿景，围绕推广自然生态保护的中心任务，积极开展生态环保科研调查、自然生态保护的推广、野生鸟类观测及环境监测等工作，以期提升“尊重自然、保护环境、持续参与”的价值观，促进自然生态保护可持续健康发展。

协会立足珠海本土，设立了重要鸟类观测点，经常性地开展鸟类观测和自然生态调查工作，建立了海量的珠海鸟类资源数据库和影像资料库；协会具备较强的科学调研能力，承接完成了多项动物资源本底调查、湿地资源保育、鸟类巡护监测及公众宣传等多项珠海市生态科研项目；协会团队编撰出版的生态科普书籍和生态图册旨在推动生态文明建设理念深入人心，为珠海市创建国家森林城市、珠海海岛鸟类保护、横琴粤澳深度合作区的生态保护与公众宣传等做了大量卓有成效的贡献，为加快生态文明建设和绿色发展尽一份绵薄之力，受到政府、学校、社会各界和广大市民的一致好评。

# Introduction of Zhuhai Bird Watching Society

The Zhuhai Bird Watching Society was established in November 2013. It is a non-profit social organization registered with the Zhuhai Civil Affairs Bureau. The society is also a group member of the Guangdong Wetland Association, the Zhuhai Association for Science and Technology, and a member of the Zhuhai Science Popularization Alliance. In April 2019, it was awarded the 5A rating in the evaluation of social organizations in Zhuhai.

The society adheres to the vision of "harmony between humans and nature" and focuses on promoting the central task of nature conservation. It actively carries out ecological and environmental research, promotes nature conservation, conducts wild bird observations, and monitors the environment. Its goal is to enhance the values of "respecting nature, protecting the environment, and continuous engagement" and promote sustainable and healthy development in nature conservation.

The society is rooted in Zhuhai and has established important bird observation points. It regularly conducts bird observations and ecological surveys, and has established a large database and image library of bird resources in Zhuhai. The society has strong scientific research capabilities and has undertaken various ecological research projects in Zhuhai, including the surveys of terrestrial vertebrate resources, wetland conservation, and bird patrol monitoring and public awareness campaigns in Zhuhai. The society has also published popular science books to increase the awareness of ecological civilization of the public. The society has made significant and effective contributions to zhuhai's initiatives such as the creation of a "National Forest City", island bird conservation in Zhuhai islands, and ecological conservation and publicity in Hengqin. It has also played a role in advancing ecological civilization and green development. The society has received unanimous praise from the government, educational institutions, various sectors of society, and the general public.

横琴二井湾湿地公园

Erjingwan Wetland Park in Hengqin

# 后记

横琴岛与澳门隔河相望。2009 年，国务院正式批准实施《横琴总体发展规划》，成立横琴新区管理委员会，拟逐步把横琴建设成为“一国两制”下探索粤港澳合作新模式的示范区，横琴岛由此跨入经济开发高速通道。发展经济绝不能以牺牲生态和环境为代价，而应在加强环境治理基础上推动经济发展。为妥善处理经济发展与环境保护的关系，有必要摸清横琴岛的动植物等自然资源的现状。横琴新区公共建设局于 2011 年 6 月至 11 月委托华南濒危动物研究所调查横琴岛陆生脊椎野生动物，共记录到鸟类 13 目 35 科 80 种。2020 年 7 月，横琴新区建设环保局（2021 年 9 月为横琴粤澳深度合作区城市规划和建设局）委托珠海市观鸟协会进行横琴全岛及洪湾的陆生野生脊椎动物资源本地调查。

珠海市观鸟协会高度重视，专门成立了项目调查队，科学设计调查方案，并邀请北京师范大学－香港浸会大学联合国际学院詹肇泰博士、鸟类学专家廖晓东教授、云南红河学院王剑博士对调查方案提出建议，同时对调查队员进行专业培训。

2020 年 7 月至 2023 年 2 月，珠海市观鸟协会共承接横琴粤澳深度合作区野生动物巡护、监测项目 4 项，分别是《珠海市横琴岛陆生野生脊椎动物资源本底调查项目（2020—2021）》、《芒洲湿地重点保护野生动物巡护和监测项目（2021—2022）》、《横琴粤澳深度合作区环岛南路海滩生物多样性资源调查项目（2022）》、《2022 年横琴粤澳深度合作区野生动物（鸟类、两爬和兽类）巡护、监测和科普宣传项目》。4 个项目均通过相关领域专家组评审通过。

生态调查队采用样线调查的方式，辅以运用红外相机、无人机等专业设备，样线遍布横琴全岛。结合历史文献资料，截止 2023 年 2 月，共记录横琴岛野生鸟类 222 种，隶属于 18 目 55 科，包括国家级重点保护野生动物 37 种。而项目调查成果更为横琴岛增加野生鸟类新记录 64 种。

项目调查数据表明，横琴岛陆生野生脊椎动物以鸟类资源最为丰富，因此，我们将其中的鸟类生态图片汇编成《横琴野生鸟类》，一是展示横琴的生态之美，二是作为向致力于横琴自然生态保护的各界人士的汇报，愿我们在人与自然相处的美好中享受经济高速发展的成果。

4 个调查项目的完成及本书的付梓，是横琴粤澳深度合作区城市规划和建设局与珠海市观鸟协会双方精诚合作的成果。项目执行期间，相关专家学者以及众多热心志愿者给与了指导与帮助，在此向他们表示诚挚的感谢！

横琴粤澳深度合作区城市规划和建设局

珠海市观鸟协会

2023 年 5 月

# Epilogue

Hengqin Island is facing Macau across a river. In 2009, the State Council officially approved the implementation of the Hengqin Overall Development Plan and established the Hengqin New Area Administrative Committee with the aim of gradually developing Hengqin into a demonstration zone for exploring the new models of Guangdong–Hong Kong–Macao cooperation under the principle of "one country, two systems." This marked the entry of Hengqin Island into a high–speed economic development.

In order to ensure the sustainable development of the economy without sacrificing the environment and ecology, it is necessary to understand the current status of the natural resources, including flora and fauna, on Hengqin Island. The Construction Bureau of Hengqin entrusted the Southern China Institute of Endangered Animals to investigate the terrestrial vertebrates in Hengqin between June and November in 2011. It was recorded total 80 species of birds in 13 orders and 35 families.

In July 2020, the Construction and Environmental Protection Bureau of Hengqin (the Urban Planning & Construction Bureau of Guangdong–Macao In–Depth Cooperation Zone in Hengqin after September 2021) commissioned the Zhuhai Bird Watching Society (ZHBWS) to conduct a complete survey of terrestrial wild vertebrate animal resources in the entire Hengqin Island and the Hongwan area.

Zhuhai Bird Watching Society has attached great importance to this initiative and has established an eco–investigation team. Dr. Tsim Siu Tai of Beijing Normal University–Hong Kong Baptist University United International College, Professor Liao Xiaodong, an ornithology expert, and Dr. Wang Jian of Honghe University in Yunnan Province were invited to provide suggestions on the survey plan and give professional training to the eco–investigation team of ZHBWS to ensure the quality and accuracy of the survey.

Between July 2020 and February 2023, the ZHBWS was entrusted 4 survey and monitoring projects altogether. They are *Baseline Survey of terrestrial vertebrates in Hengqin (2020–2021)*, *Monitoring of the National Protected Wild Animals in Mangzhou Wetland Park (2021–2022)*, *Survey of Biodiversity Resources of HuandaoNan Road in Hengqin (2022)*, and *Monitoring of terrestrial vertebrates in Hengqin (2022)*. They were all reviewed and approved by the expert panel in the relevant area. The eco–investigation team conducted the survey and monitoring by using the line transect, camera–trapping, and drones to cover the overall Hengqin Island. Combined with the historical data, 222 bird species of 18 orders and 55 families were recorded in Hengqin by February 2023, including 37 species of the national protected wild animals. In addition, the project outcomes included 64 new bird species records in Hengqin.

The project outcomes indicate that Hengqin Island is rich in bird resources. Therefore, the ecological bird photographs have been compiled into this book titled to showcase the ecological beauty in Hengqin and serve as a report to individuals who dedicated to the protection of Hengqin's natural ecology. May we enjoy the fruits of rapid economic development while cherishing the harmonious between human beings and nature!

The completion of these four projects and the publication of this book would not have been possible without the genuine cooperation between the Urban Planning & Construction Bureau of Guangdong–Macao In–Depth Cooperation Zone in Hengqin and Zhuhai Bird Watching Society. In addition, sincere gratitude would be given to the helps and contributions from those professors, experts and volunteers.

Urban Planning & Construction Bureau of Guangdong-Macao In-Depth
Cooperation Zone in Hengqin
Zhuhai Bird Watching Society
May. 2023